Induction, Recursion, and Programming

Induction, Recursion, and Programming

Mitchell Wand
Indiana University

North Holland
New York • Oxford

Elsevier North Holland, Inc.
52 Vanderbilt Avenue, New York, New York 10017

Sole distributors outside the United States and Canada:

Elsevier Science Publishers, B.V.
P.O. Box 211, 1000 AE Amsterdam, The Netherlands

Second printing, 1981

Quote on p. xiii is from Edsger W. Dijkstra, *A Discipline of Programming*, Prentice-Hall, Englewood Cliffs NJ, 1976.

Library of Congress Cataloging in Publication Data

Wand, Mitchell.
Induction, recursion, and programming.

Bibliography: p.
Includes index.
1. Electronic digital computers—Programming. 2. Induction (Mathematics) 3. Recursion theory. I. Title.
QA76.6.W34 001.6'42 79-25360
ISBN 0-444-00322-3

Desk Editor Louise Schreiber
Design Edmée Froment
Art rendered by Vantage Art, Inc.
Cover Design Paul Agule Design
Production Manager Joanne Jay
Compositor Science Typographers, Inc.
Printer Haddon Craftsmen

Manufactured in the United States of America

For Barbara

Contents

To the Instructor

This text is intended to be used in a one-semester course in discrete structures, similar but not identical to course B3 of ACM Curriculum 68 [ACM 68]. The conventional approach to B3 is "everything the computer science major ought to know about mathematics." In practice it is therefore not a computer science course, but a mathematics course whose connections with computer science are tenuous at best. The kind of "theory" a student should be learning at this level is not the usual collection of definitions and lemmas but something which is immediately useful to him. The function of theory is to help the scientist (or student) understand a complex real-world situation by explaining it in a more tractable form.

We have therefore selected topics which have immediate relevance to the problem of *programming*. The student taking a course in discrete structures is just starting to see modern ways of doing programming in his courses in programming languages and data structures. She[1] is surely in need of some theory to give some coherence to these new ideas. The last ten years have seen the development of precisely this theory. Our goal in this text is to give the student enough background to appreciate and apply this theory to the programming task.

If the past decade's work has one central thesis, it is that a program is a mathematical object, worthy of study by mathematical methods. Since the mathematics we are presenting concerns formal objects (programs) rather than the more conventional informal objects, the presentation sometimes has a somewhat metamathematical flavor. We have tried to leaven this with useful examples. Furthermore, the mathematical topics covered include a large portion of those included in a more conventional discrete structures course.

Chapter 1 is a review of basic concepts and a brief introduction to graph

[1] I have alternated the use of "he" and "she" to avoid sexism.

theory. Chapter 2 is devoted to mathematical induction. Here we begin our concern with program correctness by using induction to prove the correctness of some simple flowchart programs. The chapter also emphasizes the use of induction as a definition tool and introduces some Backus–Naur-type definitions.

Chapter 3 is devoted to a study of a particular problem which is of importance to the student at this stage of his career: recursion. We develop the semantics of recursive schemes in the style of [Knuth 68], using first-order structures as a model for data types. From this extended case study the student should learn "morals" that will be useful to her:

the distinction between an object and its name;

the central role of induction;

mathematics as modeling tool.

In addition he should learn some more specific things:

call by name vs. call by value;

techniques for proving the correctness of recursive programs.

Chapter 4 is a brief excursion into computability theory, presenting a universal machine (alias the interpreter) and the unsolvability of the halting problem. Chapter 5 uses the same framework to develop the propositional and predicate calculi. This is a prelude to Chapter 6, in which Hoare's theory of program proving is presented. This chapter emphasizes the distinction between truth and provability for partial correctness assertions. It also emphasizes the use of loop invariants as a program-writing tool.

We assume the student has had at least one mathematics course beyond high school algebra (almost any will do, so long as it covers sets, functions, and the idea of a proof) and at least one, preferably two, semesters of programming experience. A "software-oriented" course would go straight through the book in one semester. A more traditional course could be fashioned by deleting Sections 3.4–4.3 and substituting more material on algebraic structures. By covering Chapters 1–3 in the first half of the term, this course can be corequisite rather than prerequisite to a course on data structures or programming languages.

This course has been taught, by me and by seven other instructors, to sophomores at Indiana University since 1973. Our thanks go out to the many hundreds of students who dealt with the text in various stages of incompletion and to the several generations of technical typists who struggled with the manuscript. Particular thanks go to Dan Friedman for his continual encouragement and enthusiasm for the project over the years.

Introduction for the Student

Consider the following pairs of things:

"DRAWING OF A BRICK"

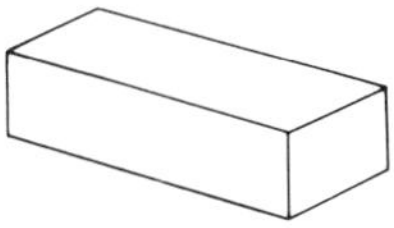

"A RECTANGLE"

```
FUNCTION F(X)
     F = X*X
END
```

$$f(x) = x^2$$

If you understand the distinction between each pair, you already understand the first major concept in this course. In each case the thing in the left-hand column is a description—a string of letters on the page. In each case the thing on the right-hand side is the *object* to which the description refers. Given a description, we use our knowledge of the world to figure out the object to which it refers. Given an object, we use our knowledge of the world to find a description for it. This distinction underlies one of the first concepts in programming: the distinction between the name of a variable (say x) and the contents of the variable (say the integer 3).

A program itself is a description—a string of letters on the page. The object corresponding to the program is the function it computes. Given a function to compute (an object) we use our knowledge of programming to find a program (a description) which computes it. Given a program, on the other hand, what is the function it computes?

Already we have seen a concept—the distinction between descriptions and objects—and a goal—to associate functions (objects) to programs (descriptions)—which will be central to this course.

Another central concept is *mathematical induction.* Induction is our major technical tool. It enables us to define infinite objects like sets and functions by a finite description. Almost everything in this text is done by induction. Unlike the description/object distinction, the concept of induction is not instilled in us from childhood. The wise student, therefore, will pause early on to understand induction thoroughly and to become familiar with it before proceeding. Familiarity breeds not contempt but comfort.

Since a program is also a finite description of an infinite object (the set of possible computations it may preform), it is not surprising that induction and programming are related. It turns out that the programming technique called *recursion* is nothing more than mathematical induction. If you understand induction now, you will see nothing odd in recursion.

Throughout this text, we emphasize the notion that a program is a mathematical object. We therefore study the mathematical properties of programs. We mathematically prove the correctness of programs. We also use mathematical techniques to help us write programs.

Chapter 1 presents some mathematical preliminaries. Most of the material in the first section or so ought to be familiar; after that we edge into material that will probably be new to you.

Chapter 2 is devoted to mathematical induction. There we do our first proof of correctness of a program. We also will see how induction is used to define sets and functions.

Chapter 3 shows how we can use these techniques to get a mathematically rigorous, unambiguous definition of a small programming language with recursion.

In Chapter 4 we consider a moderate-sized programming problem using the language of Chapter 3. We prove that there are tasks which cannot be performed by any finitely long program, no matter how clever.

In Chapter 5, we take up the language of logic, which is a language for describing *facts,* just as a programming language is a language for describing algorithms.

In Chapter 6, we merge descriptions of facts and descriptions of algorithms to develop some general techniques for proving facts about programs. The key idea is that of a *loop invariant,* which enables us to write down in a concrete way some intuition about why a program works. In the last sections of Chapter 6, we show how these ideas can be used to help *write* programs.

We have attempted to be as informal as the subject matter will allow. I have used the footnotes for a variety of purposes: to give alternative explanations, to give supplementary information, to reassure (or occasionally caution) the student at critical points, and to make connections, both with the knowledge the student may bring to the course and with more advanced fields of research.

After having devoted a considerable number of years of my scientific life to clarifying the programmer's task, with the aim of making it intellectually better manageable, I found this effort at clarification to my amazement (and annoyance) repeatedly rewarded by the accusation that 'I had made programming difficult'. But the difficulty has always been there, and only by making it visible can we hope to become able to design programs with a high confidence level, rather than 'smearing code', i.e., producing texts with the status of hardly supported conjectures that wait to be killed by the first counterexample.

Edsger W. Dijkstra

"But 'glory' doesn't mean 'a nice knock-down argument,' " Alice objected.

"When *I* use a word," Humpty Dumpty said, in a rather scornful tone, "it means just what I choose it to mean—neither more nor less."

"The question is," said Alice, "whether you *can* make words mean so many different things."

"The question is," said Humpty Dumpty, "which is to be master—that's all."

Lewis Carroll

Induction, Recursion, and Programming

1 Sets, Graphs, and Relations

1.1 SETS, RELATIONS, AND FUNCTIONS

We assume you have some intuitive idea of what a *set* is. A set is a "collection" of "things."[1] The "things" in a set are called its *members* or *elements*. If A is a set and x is an element of A, we write

$$x \in A$$

If x is not a member of A, we write $x \notin A$. We will often put a slash through a symbol to indicate "it is not the case that...." We will use braces to signify sets. Thus

$$\{\underline{a}, \underline{b}, \underline{c}\}$$

denotes the set whose members are just the letters $\underline{a}, \underline{b}, \underline{c}$. We will write

$$\{x \mid x \text{ has property } P\}$$

to denote the set of all things x that have property P.[2]

A set is determined by its members. That is, if A and B are sets, $A = B$ if and only if the following statement is true:

$$\text{for every} \quad x, \qquad x \in A \qquad \text{if and only if} \quad x \in B$$

[1]For this sentence to make sense, of course, we would have to make clear what is meant by "collection" and "thing." Since these terms have been debated by philosophers for thousands of years, we would be presumptuous to try and settle them. So a set is whatever you think it is, provided your ideas are consistent with the examples in this section. But don't worry: Your notions of "collection" and "thing" are probably not much different from everybody else's.

[2]This sentence gets us into the same trouble as the one in the previous note: What is a "property"? Again, the best answer is that a property is just what you always thought it was (but were afraid to say): something like "is red," "is green," or "is divisible by 3." Actually, things are not as fuzzy as all that: The subjects of *mathematical logic* and *set theory* have evolved since 1890 and have gone far toward making some sense out of these notions. But for the present your intuitive notion is quite good enough.

Thus $\{\underline{a},\underline{b}\}=\{\underline{b},\underline{a}\}=\{\underline{a},\underline{b},\underline{b}\}$, since all three sets have precisely the same members.

We say A is a subset of B iff[3] every member of A is a member of B. We write $A \subseteq B$. Note that $A \nsubseteq B$ iff there is some x such that $x \in A$ but $x \notin B$. For every A, $A \subseteq A$. Examples of sets follow:

EXAMPLE 1. The set with no members[4] is called the *empty set* and is denoted $\varnothing$. For every set A, $\varnothing \subseteq A$, since you cannot name a member of $\varnothing$ that is not a member of A.

EXAMPLE 2. If A and B are sets, the *product* of A and B, denoted $A \times B$, is the set of all ordered pairs (a,b) such that $a \in A$ and $b \in B$. We write

$$A \times B = \{(a,b) | a \in A \text{ and } b \in B\}$$

For example, if $A=\{\underline{a},\underline{b}\}$ and $B=\{\underline{a},\underline{b},\underline{c}\}$, then

$$A \times B = \{(\underline{a},\underline{a}),(\underline{a},\underline{b}),(\underline{a},\underline{c}),(\underline{b},\underline{a}),(\underline{b},\underline{b}),(\underline{b},\underline{c})\}$$

Note that $(\underline{a},\underline{b})$ and $(\underline{b},\underline{a})$ are distinct members of $A \times B$: these are *ordered* pairs and $(\underline{a},\underline{b}) \neq (\underline{b},\underline{a})$.[5]

Similarly, we can form the product of several sets. The product of three sets is a set of ordered triples, and so on. Thus:

$$A \times B \times C = \{(a,b,c) | a \in A \,\&\, b \in B \,\&\, c \in C\}$$

A useful special case is the product of a set with itself. For instance, $A \times A = \{(a_1,a_2) | a_1 \in A \,\&\, a_2 \in A\}$. We call this set A^2 (just as 5×5 is called 5^2). Similarly, for any natural number n, we define the nth power of A to

[3] We will use the abbreviation "iff" for "if and only if."

[4] Don't panic! (If the empty set didn't bother you, don't even read this note.) Surely, you can have sets with 1,2,3,... members, so why not a set with 0 members? This is one thing your "intuitive notion" of set may have missed. It turns out to be useful to have around: The set of all integers greater than 4 and less than 3 is certainly a "collection of things"—it just happens not to have anything in it. Like a museum without any paintings, the walls ("{" and "}") are there; there just isn't anything hung inside. If you still cannot make yourself believe in the empty set, just pretend $\varnothing$ is a special symbol and redefine a mathematical set as "anything I thought was a set, plus this special guy $\varnothing$."

> "What's the good of Mercator's North Poles and Equators
> Tropic Zones and Meridian Lines?"
> So the Bellman would cry: and the crew would reply
> "They are merely conventional signs!"
>
> Lewis Carroll

[5] Here is another source of possible confusion: We write $(x,y) \in S$ to mean that the ordered pair (x,y) is one member of the set S. We also write "$x,y \in S$" as an abbreviation for "$x \in S$ and $y \in S$." This is terrible, but it is standard mathematical shorthand, so we all have to live with it.

be a set of n-tuples[6]

$$A^n = \underbrace{A \times \cdots \times A}_{n \text{ times}} = \{(a_1,\ldots,a_n) | a_1,\ldots,a_n \in A\}$$

EXAMPLE 3. If A and B are sets, their *union*, denoted $A \cup B$, is the set

$$\{x | x \in A \text{ or } x \in B\}$$

In Example 2, $A \cup B = \{\underline{a}, \underline{b}, \underline{c}\}$. Note that $A \subseteq A \cup B$ and $B \subseteq A \cup B$.

EXAMPLE 4. If A and B are sets, their *intersection*, denoted $A \cap B$, is the set

$$\{x | x \in A \text{ and } x \in B\}$$

With the A and B of Example 2, $A \cap B = \{\underline{a}, \underline{b}\}$. If $A \cap B = \varnothing$, we say A and B are *disjoint*.

EXAMPLE 5. If A and B are two sets, their *difference*, denoted $A - B$, is the set

$$\{x | x \in A \text{ and } x \notin B\}$$

EXAMPLE 6. If A is a set, its *power set*, denoted $P(A)$, is the set

$$\{B | B \subseteq A\}$$

Notice that $P(A)$ is a set whose members are sets. This is often confusing, but it need not be: A set was a collection of "things" and a set is certainly a "thing"! If $A = \{\underline{a}, \underline{b}, \underline{c}\}$, then

$$P(A) = \{\varnothing, \{\underline{a}\}, \{\underline{b}\}, \{\underline{c}\}, \{\underline{a}, \underline{b}\}, \{\underline{a}, \underline{c}\}, \{\underline{b}, \underline{c}\}, \{\underline{a}, \underline{b}, \underline{c}\}\}$$

EXAMPLE 7. The *natural numbers* are a set, denoted N:

$$N = \{1, 2, 3, \ldots\}$$

EXAMPLE 8. The *integers* are a set, denoted Z:

$$Z = \{\ldots, -3, -2, -1, 0, 1, 2, 3, \ldots\}$$

EXAMPLE 9. The *nonnegative integers* are a set, denoted ω (the Greek letter *omega*)[7]:

$$\omega = \{0, 1, 2, \ldots\}$$

[6]An *n-tuple* is like an ordered pair or ordered triple, except that it has n members. So an ordered pair is a 2-tuple, an ordered triple is a 3-tuple, and an ordered quadruple is a 4-tuple.

[7]Please, again: don't panic. Is "ω" much worse than "&"? We will be very sparing in our use of Greek letters (and there is a table of them in the back of the book). Don't say "curly w," say "omega," and it will soon be as easy as π.

EXAMPLE 10. Let V be any set. Think of V as a set of letters. Then V^+ is the set of all possible words made up of letters from V. For example, if $V=\{\underline{a},\underline{b},\underline{c}\}$,

$$V^+=\{\underline{\mathrm{a}},\underline{\mathrm{b}},\underline{\mathrm{c}},\underline{\mathrm{aa}},\underline{\mathrm{ab}},\underline{\mathrm{ac}},\underline{\mathrm{ba}},\underline{\mathrm{bb}},\underline{\mathrm{bc}},\underline{\mathrm{ca}},\underline{\mathrm{cb}},\underline{\mathrm{cc}},\underline{\mathrm{aaa}},\underline{\mathrm{aab}},\underline{\mathrm{aac}},$$
$$\underline{\mathrm{aba}},\underline{\mathrm{abb}},\underline{\mathrm{abc}},\underline{\mathrm{aca}},\underline{\mathrm{acb}},\underline{\mathrm{acc}},\underline{\mathrm{baa}},\underline{\mathrm{bab}},\underline{\mathrm{bac}},\underline{\mathrm{bba}},\ldots,\underline{\mathrm{ccc}},$$
$$\underline{\mathrm{aaaa}},\ldots\}$$

We refer to the members of V^+ as *words* or *strings*. Sets like V^+ are very important in computer science. For example, let V be the set of legal card punches, i.e., $V=\{\underline{0},\ldots,\underline{9},\underline{A},\ldots,\underline{Z},\underline{+},\underline{-},\underline{.},\ldots,\underline{,},\underline{\$}\}$. We use underlines to emphasize the fact that these are merely symbols (like cardpunch codes) and do not have any special meaning. In this way we will not confuse the arithmetic operation + and the symbol $\underline{+}$. We sometimes call "mere symbols" *formal symbols*, which is their proper mathematical name.

So, for example, we are concerned with sets like

$$\{w\in V^+ \mid w \text{ is a legal FORTRAN arithmetic expression}\}$$
$$\{w\in V^+ \mid w \text{ is a legal ALGOL program}\}$$[8]

EXAMPLE 11. Another interesting set is the set $\{T,F\}$, or {TRUE, FALSE}. We do not have a special name for this set.

Some types of sets have special names:

Definition. If A and B are sets, a *relation from A to B* is a subset of $A\times B$.

We will see many examples of relations in later sections.

Definition. A *function from A to B* is a relation f from A to B such that for every $a\in A$ there is one and only one $b\in B$ such that $(a,b)\in f$. If for every $a\in A$ there is at most one $b\in B$ such that $(a,b)\in f$, we say f is a *partial function*.[9] The set A is called the *domain* of f and the set B is called the *range* of f.

[8]No, you need not know ALGOL for this course. But we hope this course will be a help in understanding ALGOL later.

[9]While you have probably seen functions before, the concept of partial function may be new to you. A function f from A to B may be thought of as a "black box." Given an input chosen from the set A, the box gives an answer from the set B, namely, the unique b such that $(a,b)\in f$. A partial function is just like a function except that for some $a\in A$ there may be *no* $b\in B$ such that $(a,b)\in f$, and so the black box gives no output. As we shall see, partial functions arise naturally in programming.

We write f: $A \to B$ to indicate that f is a function from A to B. If $(a,b) \in f$, we say b is the *value* of f at a, and write $f(a) = b$. This, of course, is nothing more than the usual notion of a function.

For example, let $A = \{\underline{a},\underline{b},\underline{c}\}$, $B = \{\underline{a},\underline{b},\underline{c},\underline{d}\}$, and $f = \{(\underline{a},\underline{b}),(\underline{b},\underline{b}),(\underline{c},\underline{d})\}$. Then f is a function from A to B, with $f(\underline{a}) = \underline{b}$, $f(\underline{b}) = \underline{b}$, and $f(\underline{c}) = \underline{d}$. The relation f is not a function from B to B, but it is a partial function from B to B, since there is no value for $f(\underline{d})$. The relation f is neither a function from A to A nor a partial function from A to A, since f is not a subset of $A \times A$.

A phrase such as "let f: $A \to B$" is like a declaration in a programming language. It says that for the purposes of the next piece of text (paragraph, proof, etc.), the symbol f denotes a function from the set A to the set B. Like any declaration, it must include all of the required information—in this case the domain and range. The question, "Is f a function?", is not meaningful unless we specify the domain and range. For example, $\{(1,2),(2,2)\}$ is a function from $\{1,2\}$ to $\{2\}$ or to $\{1,2,3\}$ but it is not a function from $\{1,2,3\}$ to $\{1,2,3\}$, since there is no value for $f(3)$. In a similar manner, we will always specify the domain and range of a relation $R \subseteq A \times B$.

We next consider some interesting classes of functions:

1. If f: $A \to B$ has the property that every $b \in B$ is a possible output of f, we say f is *surjective* or *onto*. More formally, we say f: $A \to B$ is surjective iff for each $b \in B$ there is an $a \in A$ such that $f(a) = b$.
2. If f: $A \to B$ always sends distinct elements of A to distinct elements of B, we say f is *one-to-one* or *injective*. More formally, we say f: $A \to B$ is injective iff for every a and $a' \in A$, if $a \neq a'$, then $f(a) \neq f(a')$. An equivalent statement is: If $f(a) = f(a')$, then $a = a'$.
3. If f: $A \to B$ is one-to-one and onto, we say f is *bijective* (or f is a *bijection*).
4. If f: $A \times B \to C$, we say f is a *two-place* function which takes its first argument from A and its second argument from B. Such an f is a relation from $A \times B$ to C, and so it is a subset of $(A \times B) \times C$. Therefore a typical element of f is $((a,b),c)$[10] and by the notation we used before, we should write $f((a,b)) = c$. But that is too cumbersome, so we write $f(a,b) = c$ instead. Similarly, we write $g(a,b,c)$ for the result of applying a three-place function, etc.
5. A function P: $A \to \{T,F\}$ is called a *predicate* on A. So for example, we have the predicate P on ω defined by

$$P(n) = \begin{cases} \text{TRUE} & \text{iff } n \text{ is odd} \\ \text{FALSE} & \text{otherwise} \end{cases}$$

[10]Remember, by our definitions f must be a set of ordered pairs, not ordered triples or anything else.

Every predicate P on A determines a subset of A as follows:

$$\{x \mid x \in A \text{ and } P(x) = \text{TRUE}\}$$

We sometimes abbreviate this by

$$\{x \in A \mid P(x)\}$$

Similarly, given f: $A \rightarrow B$, we can define a subset of B (called the *image* of f) by $\{f(x) \mid x \in A\}$.

The integers have a number of interesting functions associated with them, for example, $+$: $Z \times Z \rightarrow Z$ (addition) and $*$: $Z \times Z \rightarrow Z$ (multiplication). There is an important operation on V^+ with which you should be familiar. Let $\circ$: $V^+ \times V^+ \rightarrow V^+$ be given as follows: $\circ(w, v)$ is the string obtained by writing down the symbols in w and then writing down the symbols in v. For example, $\circ(\underline{\text{ab}}, \underline{\text{bc}}) = \underline{\text{abbc}}$. This operation is called *concatenation*, and we often write wv for $\circ(w, v)$. You can readily see that for any strings w, v, and x, $(wv)x = w(vx)$. So, just as is the case for integer addition, we delete the parentheses and write wvx. But note that, unlike integer addition, it is false that $wv = vw$: $\underline{\text{ab}} \neq \underline{\text{ba}}$. Many important facts about programming languages are expressible in terms of concatenation. For example, if v and w are legal FORTRAN expressions, then so is $v \underline{+} w$.[11]

This discussion of concatenation brings up another bit of notation. Addition is a two-place function $+$: $Z \times Z \rightarrow Z$. So, according to our previous notation, we should write things such as

$$+(3,5) = 8$$

Nevertheless, we are used to writing

$$3 + 5 = 8$$

"$+(3,5)$" is called *prefix* notation and "$3+5$" is called *infix* notation. Infix is the usual notation for two-place arithmetic operators, and we shall see that it is convenient for other two-place operators as well. For concatenation, we have gone one step further and written wv instead of $w \circ v$ or $\circ(w, v)$. This seems quite reasonable, since the value of the concatenation wv is just the result of writing down the string which is the value of x *next to* the string which is the value of v.

[11]Stop now and make sure you understand this expression. Why are v and w not underlined? Because they are *variables* that *stand for* strings. So if v stands for the string $\underline{3+2}$ and w stands for $\underline{X * Y}$, then $v \underline{+} w$ stands for (denotes) $\underline{3+2+X*Y}$. Note that if v stands for $\underline{3}$ and w stands for $\underline{2}$, then $v \underline{+} w$ stands for $\underline{3+2}$, not $\underline{5}$! $v \underline{+} w$ is a description of the *object* $\underline{3+2}$; since $\underline{3+2}$ is also a string, it may also be the description of some other *object* (perhaps, if we are lucky, the integer 5). Compare this with the discussion in the Introduction for the Student.

Two-place functions are very common and may have several special properties:

If $\circledast: A \times A \to B$, we say $\circledast$ is *commutative* iff for all $a, a' \in A$, $a \circledast a' = a' \circledast a$.

If $\circledast: A \times A \to A$, we say $\circledast$ is *associative* iff for all $a, b, c \in A$, $a \circledast (b \circledast c) = (a \circledast b) \circledast c$.

If $\circledast: A \times A \to A$, we say $e \in A$ is an *identity* for $\circledast$ iff for all $a \in A$, $a \circledast e = e \circledast a = a$.

Integer addition is associative and commutative with zero as an identity. Integer multiplication is associative and commutative with 1 as an identity. Integer subtraction is *not* commutative or associative, and has no identity. Concatenation $V^+ \times V^+ \to V^+$ is associative but not commutative.

If $\circledast: A \times A \to A$ is associative, then we can write $a \circledast b \circledast c$ without needing parentheses to indicate the order of operations. Thus we can "take the $\circledast$" of 2,3,4, or any number of elements of A. What does one get by "taking the $\circledast$" of 0 elements of A? This should be an element of A (let us call it a_0) such that taking $\circledast$ of a_0 and any other $a \in A$ gives a. Thus a_0 must be an identity of $\circledast$.

As an example, look at the familiar program for taking the sum of the inputs:

```
sum←0;
while not end-of-file do begin
      read(a);
      sum←sum + a
      end
```

We can generalize this to a program for "taking the $\circledast$" of the inputs, where $\circledast$ is an associative function with identity e:

```
s←e;
while not end-of-file do begin
      read (a);
      s←s ⊛ a
      end
```

In the sum program, whenever we execute the end-of-file test, *sum* contains the sum of the inputs read so far. In the second program, whenever we execute the end-of-file test, s contains "the $\circledast$ of" the inputs read so far. By initializing s with the identity, these statements are guaranteed to be true the first time the loop is entered and the loop body makes sure they stay true. Statements such as these are called loop *invariants*, and we shall see that they will play an important role in our analysis of computer programs.

EXERCISES 1.1

In the following exercises, let

$$A=\{1,2\}$$
$$B=\{2,3,4\}$$
$$C=\varnothing$$
$$V=\{\underline{a},\underline{b},\underline{c}\}$$
$$W=\{\underline{b},\underline{c},\underline{d}\}$$

1. List the following sets:

(a) $A\times B$
(b) $A\cup B$
(c) $A\cap B$
(d) $B\times C$
(e) $B\cup C$
(f) $B\cap C$
(g) B^2
(h) A^3
(i) $P(A)$
(j) $P(C)$
(k) $P(\{1\})$

2. Is $A\subseteq B$, $B\subseteq A$? Which of A, B, and C are subsets of which others?

3. List $\{vw|v\in V, w\in W\}$.

4. Which of the following relations from A to B are functions?

(a) $\{(1,3),(2,4)\}$
(b) $\{(1,3),(1,4)\}$
(c) $\{(1,3),(1,3)\}$
(d) $\{(1,3),(2,5)\}$
(e) $\{(2,2),(1,4)\}$

5. Is $\{(1,2),(2,3)\}$ a function

(a) from $\{(1,2)\}$ to $\{(2,3)\}$?
(b) from N to N?
(c) from $\{1,2\}$ to N?
(d) from $\{1,2,3\}$ to $\{2,3\}$?
(e) from $\{1,2,3\}$ to $\{1,2,3\}$?

6. List a relation that is an injective function from A to B. List a relation that is a surjective function from B to A.

7. List two bijections from B to B.

8. *Prove*: If X and Y are sets, then $X=Y$ iff $X\subseteq Y$ and $Y\subseteq X$.

9. *Prove*: If f: $X \to Y$ is a bijection, then there is a bijection g: $Y \to X$.

10. Let f: $A^2 \to A$ be given by the following table:

x	y	$f(x,y)$
1	1	1
1	2	2
2	1	2
2	2	2

Show that f is commutative and associative. What is the identity of f?

11. *Prove*: If f: $X \times X \to X$, and e and e' are both identities of f, then $e = e'$.

1.2 RELATIONS AS GRAPHS

In this section and the next, we shall see some other ways to look at sets with special properties.

Recall that a relation R from A to B is a subset of $A \times B$. We can represent R as a diagram as follows: We write down the elements of A in a line and the elements of B in another line, and for each $(a,b) \in R$, we draw an arrow from the point labeled a to the point labeled b. This is called the *bipartite graph* representation of R (see Figure 1.2.1).

If A and B are the same set, then we can use another representation of R that is much more interesting: the *directed graph* representation. To represent $R \subseteq A \times A$, we write down a diagram with one point for each element of A; for each $(x,y) \in R$, we write down an arrow from the point labeled x to the point labeled y. We sometimes call the "points" *nodes* or *vertices*; we also call the arrows *edges*. We then say "there is an arrow from x to y." This is quite different from saying, "there is an arrow from y to x" (see Figure 1.2.2).

If x is a node, the number of arrows pointing to x is called its *in-degree*; the number of arrows emanating from x is called its *out-degree*. Notice that the directed graph depends on its domain and range (see Exercise 1.2.2).

Directed graphs are important in the study of data structures[12]; we introduce them now because there are several special properties of relations that have nice pictures in the directed graph representation.

[12]The term *data structures* refers to techniques for organizing data in memory in ways that reflect the relationships present in the data. The simplest data structure is the array: To keep a table of student grades one might keep a 2-dimensional array GRADES, in which GRADES $[i,j]$ contains the grade of the ith student on the jth exam. In Section 1.5 we shall present some similar techniques for representing graphs; we shall see how the resulting data structure itself begins to look like a graph. The definitive treatise on data structures is [Knuth 68].

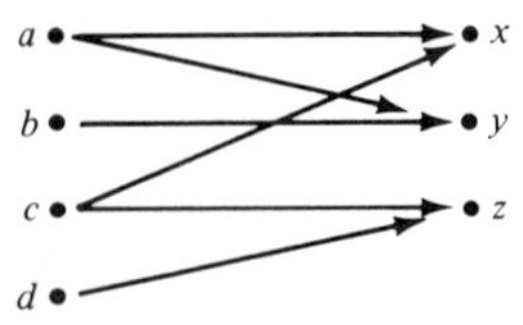

FIGURE 1.2.1 Bipartite graph representation of $R=\{(a, x),(a,y),(b,y),(c,x),(c,z),(d,z)\}$.

1. We say a relation R is *reflexive* iff for every $a \in A$, the pair $(a,a) \in R$. This means that if we draw the directed graph corresponding to R, there will be a "self-loop" on every node (see Figure 1.2.3).
2. We say R is *symmetric* iff $(x,y) \in R$, then $(y,x) \in R$. In the graph, this says that if there is an arrow from x to y, then there is an arrow in the other direction (see Figure 1.2.4). Note that in Figure 1.2.4(ii) there is an arrow from a to a. Note that the reverse arrow is also there: in fact, it is the same arrow!

   If R is symmetric, then we can draw a third representation of R. Since for every arrow there is a reverse arrow, we can delete the arrowheads entirely; then a line from one point to another stands for a pair of arrows in opposite directions. We call this the *undirected graph* representation (see Figure 1.2.5).
3. We say a relation R is *transitive* iff for every $(x,y) \in R$ and $(y,z) \in R$, $(x,z) \in R$. In terms of the directed graph, if there are ever two arrows "in sequence," then there is a *single* arrow from the tail of the first to the head of the second (see Figures 1.2.6 and 1.2.7).

FIGURE 1.2.2 Some directed graphs. For the top figure, $R=\{(a,b),(b,c),(b,d),(b,a),(c,c),(d,a)\}$, $A=\{a,b,c,d\}$. For the bottom figure, $R=\{(a,a),(a,b),(a,c),(b,c),(d,d)\}$, $A=\{a,b,c,d\}$.

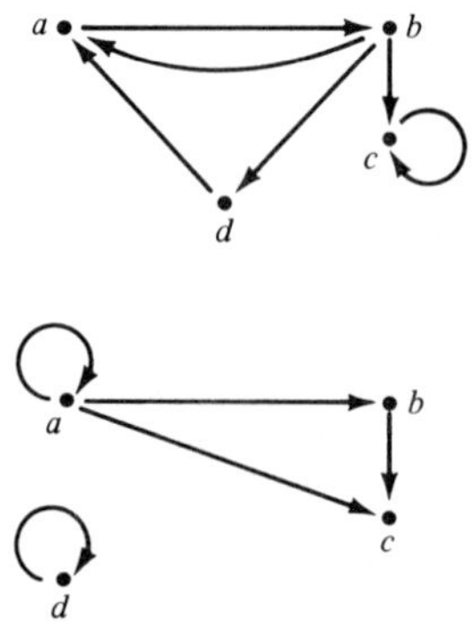

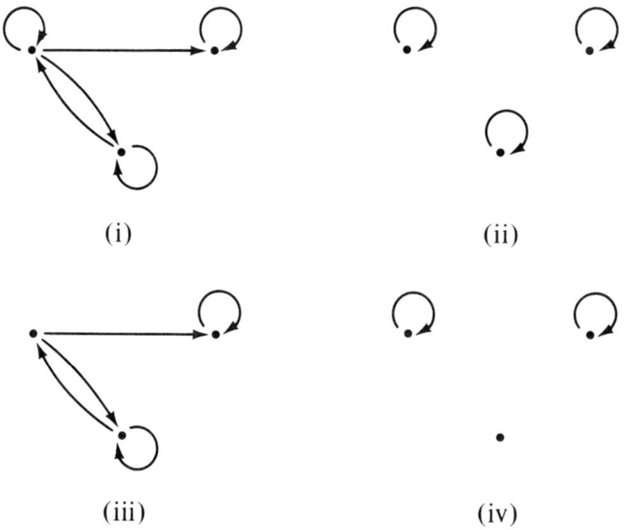

FIGURE 1.2.3 Top figures are reflexive, bottom figures not reflexive.

Let R be transitive, and (a_1,a_2), (a_2,a_3), and (a_3,a_4) all be arrows in R. By transitivity, $(a_1,a_3)\in R$, so, applying transitivity to (a_1,a_3) and (a_3,a_4), the arrow $(a_1,a_4)\in R$. Clearly, this argument can be extended to any chain of arrows. We now seek to formalize the notion of a "chain of arrows."

Definition. Let $R\subseteq A\times A$ be a relation. A *path* from a to b in R is a sequence

$$\langle a_0,a_1,\ldots,a_n\rangle$$

such that

(i) $n\geqslant 1$
(ii) for each i, $a_i\in A$
(iii) $a_0=a$, $a_n=b$
(iv) for each i such that $0\leqslant i<n$, $(a_i,a_{i+1})\in R$

If $a_0=a_n$, we call the path a *cycle*. We say n is the *length* of the path. A graph which has no cycles is said to be *acyclic*.

Proposition. R is transitive iff whenever there is a path from a to b in R, then there is an edge $(a,b)\in R$.

PROOF. ($\Leftarrow$) Assume that if there is a path from a to b in R, then there is an arrow from a to b in R. We would like to show that R is transitive, that is, if $(a_1,a_2)\in R$ and $(a_2,a_3)\in R$, then $(a_1,a_3)\in R$. If $(a_1,a_2)\in R$ and (a_2,a_3)

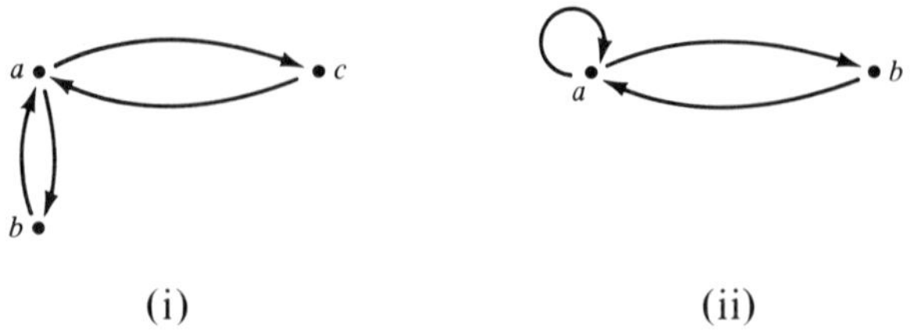

FIGURE 1.2.4 Some symmetric graphs.

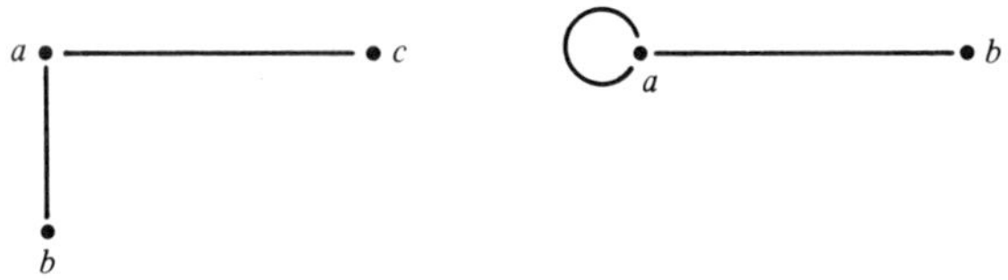

FIGURE 1.2.5 Undirected graphs of the relations of Figure 1.2.4. For the left figure, $R = \{(a,b),(b,a),(a,c),(c,a)\}$. For the right figure, $R = \{(a,a),(a,b),(b,a)\}$.

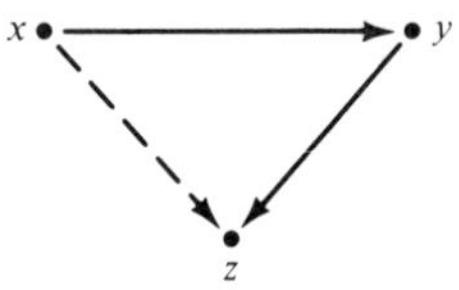

FIGURE 1.2.6 Condition for transitivity.

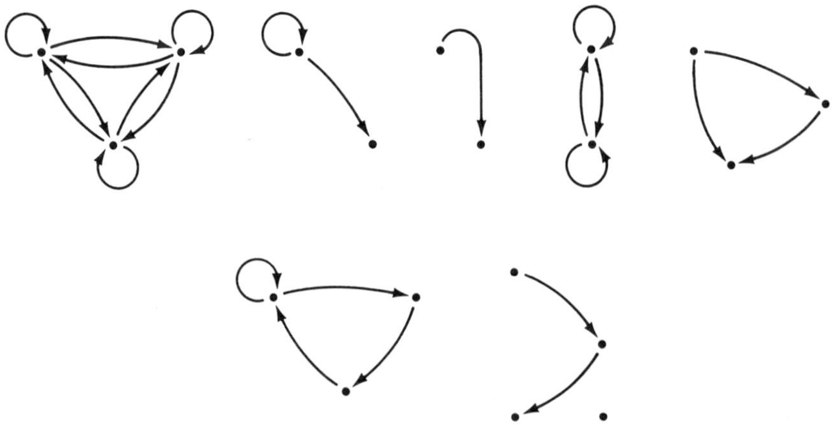

FIGURE 1.2.7 Examples of transitivity. Top figures are transitive, bottom figures not transitive.

$\in R$, then $\langle a_1, a_2, a_3 \rangle$ is a path from a_1 to a_3 in R. By the assumption, since there is a path from a_1 to a_3 in R, there is an arrow from a_1 to a_3 in R, which is just what we need to show that R is transitive.

($\Rightarrow$) Assume that $\langle a_0, \ldots, a_n \rangle$ is a path in R, and that R is transitive. We shall show that for each i such that $1 \leqslant i \leqslant n$, there is an arrow $(a_0, a_i) \in R$. We shall do this by giving an "algorithm" that, starting with the arrow (a_0, a_1), builds up an arrow (a_0, a_n) by successive applications of transitivity. Imagine we have already built the arrow $(a_0, a_i) \in R$. Since $\langle a_0, \ldots, a_i, a_{i+1}, \ldots, a_n \rangle$ is a path in R, we know that $(a_i, a_{i+1}) \in R$. Now transitivity requires that there be an arrow $(a_0, a_{i+1}) \in R$. We repeat this "extension" until we reach a_n. □

EXERCISES 1.2

1. Let

$$A = \{a, b, c, d, e\}$$
$$R = \{(a,b), (a,c), (b,a), (c,a), (c,d), (c,e), (d,c), (e,c)\}$$

   (a) Draw the bipartite graph representation of R.
   (b) Draw the directed graph representation of R.
   (c) Is R symmetric? reflexive? transitive?

2. Let $R = \{(a,a)\}$, $A = \{a\}$, and $B = \{a,b\}$. Is $R \subseteq A \times A$ reflexive? Is $R \subseteq B \times B$ reflexive? Draw both relations as bipartite graphs and directed graphs.

3. Consider the following undirected graph.

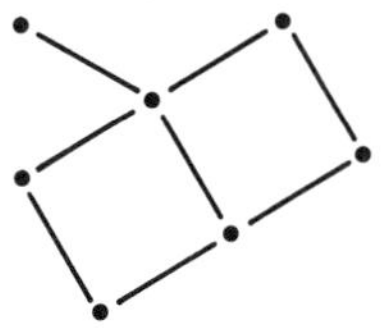

   Represent this graph as a relation.

4. Draw a directed graph that is symmetric and transitive, but not reflexive.

5. A relation that is reflexive, symmetric, and transitive is called an *equivalence* relation. Give a description of the directed graph of an equivalence relation.

6. Let $A = \{a\}$, $B = \{a, b\}$.

   (a) List all the relations $R \subseteq A \times A$.
   (b) List all the relations $R \subseteq B \times B$.
   (c) Of the relations in (a) and (b), which are reflexive? symmetric? transitive?

1.3 EQUIVALENCE RELATIONS

Relations that are reflexive, symmetric, and transitive arise so often in mathematics that they have been given a name:

Definition. A relation that is reflexive, symmetric, and transitive is called an *equivalence relation.*

Figure 1.3.1 shows the directed graph picture of a typical equivalence relation.

The directed graph picture shows that the nodes are divided into "islands." Within each island, there is an arrow from any node to any node in the same island, but between different islands there are no arrows. It is easy to see why there are no arrows between islands: If we added an arrow from some node in one island to some node in another island, then in order to make the relation transitive, we would have to add arrows between all the nodes in the two islands. These "islands" are called *equivalence classes*:

Definition. Let $R \subseteq A \times A$ be an equivalence relation, and let $a \in A$. The *equivalence class of a under R* (we write $[a]_R$) is defined as $\{a' \in A | (a,a') \in R\}$.

When we can determine R from context, we delete the "under R" and write just $[a]$. The "islands" property may be expressed as follows:

Theorem 1.3.1. *If R is an equivalence relation on A, and $a,b \in A$, then*

(i) *if $(a,b) \in R$, then $[a]=[b]$*
(ii) *if $(a,b) \notin R$, then $[a] \cap [b] = \varnothing$*

FIGURE 1.3.1 A typical equivalence relation.

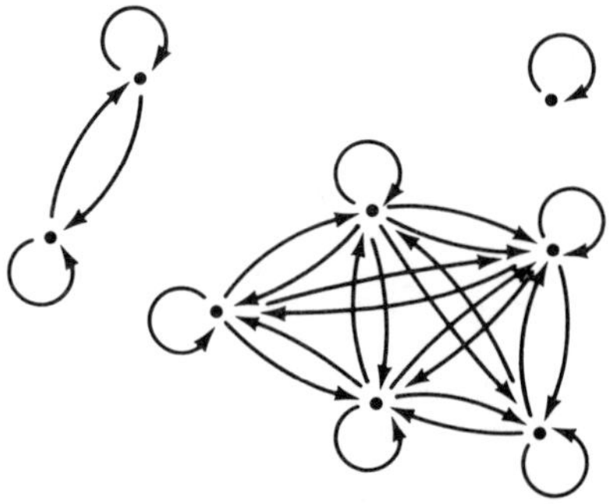

PROOF. (i) Assume $(a,b)\in R$. We shall show $[b]\subseteq[a]$ and $[a]\subseteq[b]$. To show $[b]\subseteq[a]$, let $b'\in[b]$. Then $(b,b')\in R$. Since $(a,b)\in R$ and $(b,b')\in R$, by transitivity, $(a,b')\in R$. Hence $b'\in[a]$. So any member of $[b]$ is also a member of $[a]$; so $[b]\subseteq[a]$.

To show $[a]\subseteq[b]$, let $a'\in[a]$. So $(a,a')\in R$. By symmetry, $(a',a)\in R$. Since $(a,b)\in R$ and R is transitive, $(a',b)\in R$. By symmetry again, $(b,a')\in R$, so $a'\in[b]$. Hence $[a]\subseteq[b]$.

(ii) We shall show that if $[a]\cap[b]\neq\varnothing$, then $(a,b)\in R$. Since it is impossible to have both $(a,b)\in R$ and $(a,b)\notin R$, we conclude that it is impossible to have both $(a,b)\notin R$ and $[a]\cap[b]\neq\varnothing$. Therefore, if $(a,b)\notin R$, then $[a]\cap[b]=\varnothing$.[13]

So assume $[a]\cap[b]\neq\varnothing$. Then there is some c such that $c\in[a]$ and $c\in[b]$. Since $c\in[a]$, we have $(a,c)\in R$, and since $c\in[b]$, we have $(b,c)\in R$. By symmetry, $(c,b)\in R$. Since $(a,c)\in R$ and $(c,b)\in R$, $(a,b)\in R$. By the argument of the previous paragraph, this completes the proof.[14] □

Corollary. *If $a,b\in A$, then either $[a]=[b]$ or $[a]\cap[b]=\varnothing$.*

PROOF. If $(a,b)\in R$, then $[a]=[b]$; if $(a,b)\notin R$, then $[a]\cap[b]=\varnothing$. □

It is often useful to consider the set of the equivalence classes:

Definition. If R is an equivalence relation on A, the *quotient set* A/R is $\{[a]_R|a\in A\}$.

Figure 1.3.2 shows an example of a quotient set.

Now, A/R is a subset of $P(A)$. An interesting question is: Which subsets of $P(A)$ are obtainable as quotient sets? That is, which subsets of $P(A)$ are equal to A/R for some R? The following definition and theorem supply the answer:

Definition. Let A be a set. A subset Π of $P(A)$ is a *partition* of A iff each $S\in\Pi$ is nonempty and for each $a\in A$, there is exactly one $S\in\Pi$ such that $a\in S$.

[13]This is a very important technique for proving things:

If P implies Q
then
not-Q implies not-P

This is the principle of *contraposition*; "not-Q implies not-P" is called the *contrapositive* of "P implies Q." The argument is very simple: If P implies Q, and Q is false, then P must be false also, since if P were true, then Q would be true. But Q cannot be both true and false. So P cannot be true. (Think about it!) We shall see in Chapter 5 how arguments like this can be made more comprehensible by using a formal language akin to a programming language [see Exercise 5.2.2(b)].

[14]We shall often not bother writing sentences like this.

$$A = \{1, 2, 3\}$$
$$R = \{(1, 1), (1, 2), (2, 1), (2, 2), (3, 3)\}$$
$$[1]_R = \{1, 2\}$$
$$[2]_R = \{1, 2\}$$
$$[3]_R = \{3\}$$
$$A/R = \{\{1, 2\}, \{3\}\}$$

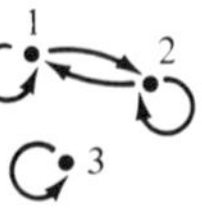

FIGURE 1.3.2

Theorem 1.3.2. *A subset Π of $P(A)$ is a partition on A iff $\Pi = A/R$ for some equivalence relation R on A.*

PROOF. ($\Leftarrow$) If R is an equivalence relation, we claim A/R is a partition. If $a \in A$, then $a \in [a]$, so there is some $S \in A/R$ such that $a \in S$. Furthermore, $[a]$ is the *only* equivalence class containing a, by the corollary to Theorem 1.3.1. Furthermore, each equivalence class is nonempty. So A/R is a partition.

($\Rightarrow$) By the definition of a partition, there exists a function $f\colon A \to \Pi$ such that for any A and $S \in \Pi$, $x \in S$ iff $S = f(x)$. Define $R = \{(a,b) | f(a) = f(b)\}$. R is easily seen to be an equivalence relation. We claim $A/R = \Pi$.

Now let $a \in A$ and $f(a) = S$. Then

$$S = \{a' | f(a') = S\} \qquad [\text{since } x \in S \text{ iff } f(x) = S]$$
$$= \{a' | f(a') = f(a)\}$$
$$= [a]_R$$

So for each a, $[a] = f(a) \in \Pi$; hence $A/R \subseteq \Pi$. Conversely, let $S \in \Pi$. By the definition of partition, S is nonempty, so choose $a \in S$. Hence $f(a) = S$, and by the previous argument, $S = [a]_R$. So $\Pi \subseteq A/R$. □

EXERCISES 1.3

List A/R for each of the following equivalence relations:

1. $A = \{x \in \omega | 1 \leqslant x \leqslant 6\}$
   $R = \{(x,y) | (x - y)$ is evenly divisible by $3\}$

2. $A = \{x \in \omega | 1 \leqslant x \leqslant 8\}$
   $R = \{(x,y) | x = 2^i k$ and $y = 2^i k'$ and k, k' are odd$\}$

3. $A = \{x \in \omega | 1 \leqslant x \leqslant 24\}$
   $R = \{(x,y) | x = 2^i 3^j k$ and $y = 2^{i'} 3^{j'} k'$ and $i + j = i' + j'$ and k, k' are not evenly divisible by 2 or 3$\}$

4. If A and B are sets and $f\colon A \to B$, define the *kernel* of f [denoted Ker(f)] to be $\{(x,y) | x,y \in A$ and $f(x) = f(y)\}$. Prove that for any f, Ker(f) is an equivalence relation.

5. *Prove:* If R is an equivalence relation on A, there exists a set B and a function $f: A \to B$ such that $R = \mathrm{Ker}(f)$. (*Hint:* Let $B = A/R$.)

6. We say two directed graphs $R \subseteq A \times A$ and $S \subseteq B \times B$ are *isomorphic* iff there exists a bijection $f: A \to B$ such that $(a,a') \in R$ iff $(b,b') \in S$.
   (a) Show that isomorphism is an equivalence relation.
   (b) For the set of all directed graphs with two nodes, draw one graph from each isomorphism equivalence class.

1.4 TREES

The graphs found in data structures are usually far from transitive. Often, there is at most a single way to get from one node to another. An important class of graphs used in data structures is the class of *trees*:

Definition. A *tree* is a finite acyclic directed graph $R \subseteq A \times A$ in which there is one node (called the *root*) with in-degree 0, and every other node has in-degree 1. A node in a tree with out-degree 0 is called a *leaf* (see Figure 1.4.1).

Again it is necessary to include A in the declaration "$R \subseteq A \times A$ is a tree" (see Exercise 1.4.2). This is one of a number of equivalent definitions in use.[15]

Theorem 1.4.1. *If $R \subseteq A \times A$ is a tree, $x \in A$, and x is not the root of R, then there is exactly one path from the root to x in R.*

PROOF. We shall first construct one path from the root to x, and then see that it is unique (see Figure 1.4.2). Since x is not the root, x has in-degree 1, so there must be a path $\langle a_1, x \rangle$. Now let us assume that we have constructed a path $\langle a_n, a_{n-1}, \ldots, x \rangle$. If a_n is the root, we are done. If a_n is not the root, then the in-degree of a_n is 1, so there must be some arrow $(a_{n+1}, a_n) \in R$. So $\langle a_{n+1}, a_n, \ldots, x \rangle$ is a path in R. How long can we make

[15]Our definition follows that of [Knuth 68, Section 2.3.4.2], who calls it an "oriented tree" or a "rooted tree" (but his arrows are in the opposite direction from ours). [Tremblay and Manohar 75] call it a "directed tree." An intimately related notion is an *undirected* or *free* tree, which is defined as follows:

$$R \subseteq A \times A \quad \text{is an } \textit{undirected tree} \text{ iff there is a tree } S \subseteq A \times A \text{ such that}$$
$$R = \{(x,y) \mid (x,y) \in S \quad \text{or} \quad (y,x) \in S\}$$

In an undirected tree, it is not possible to determine the root (see Figure 1.4.5, in which two different trees S give the same undirected tree R). Indeed, given an undirected tree, one can choose *any* node to be the root: Imagine picking up the undirected graph by the desired node and shaking it until the edges untangle themselves. The trees which arise in the study of programming languages and data structures usually seem to have natural roots, which is why we chose the definition we did.

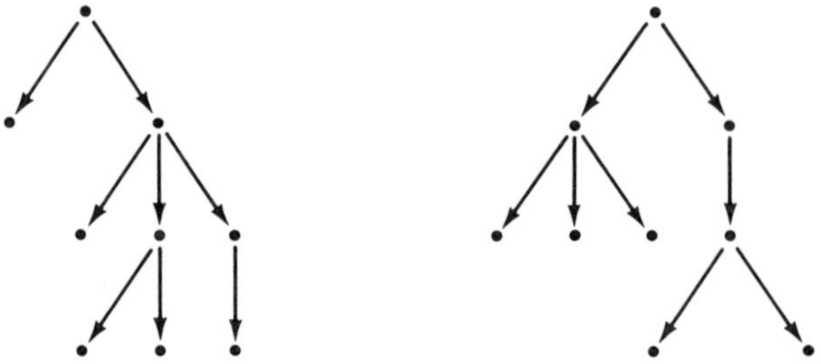

FIGURE 1.4.1 Two trees.

this path? If the path in R were longer than the number of nodes in the graph, then some node would have to be repeated. If some node were repeated, then R would have a cycle, which would imply that R was not a tree. So every path in the tree R (and our path in particular) must have length smaller than the number of nodes. Therefore, our algorithm must eventually halt. But the only way our algorithm for building paths to x can halt is by finding an a_{n+1} which is the root. So the algorithm must halt with a path from the root to x.

FIGURE 1.4.2

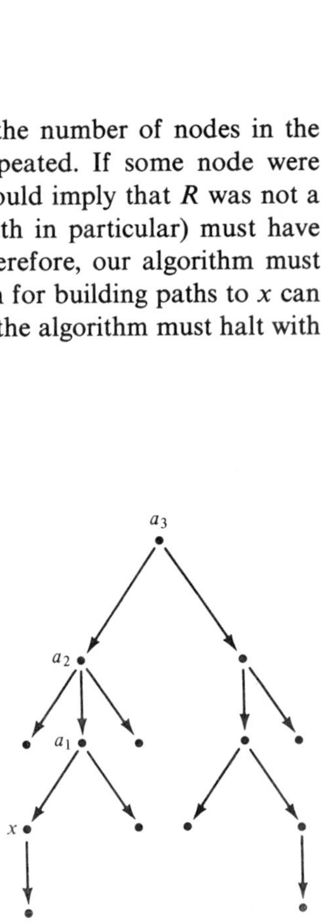

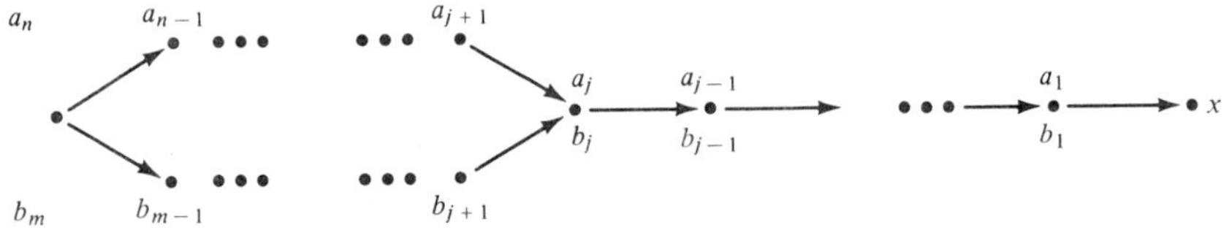

Assume that there are two paths:

$$\langle a_n, a_{n-1}, \ldots, x\rangle \quad \text{and} \quad \langle b_m, b_{m-1}, \ldots, b_1, x\rangle$$

with $a_n = b_m =$ the root. Then at some point the paths must converge: There is an integer j such that $a_{j+1} \neq b_{j+1}$, but $a_j = b_j$, $a_{j-1} = b_{j-1}, \ldots, a_1 = b_1$ (see Figure 1.4.2). So $(\quad, a_j) \in R$ and $(b_{j+1}, a_j) \in R$. Therefore a_j must have in-degree of at least 2, which contradicts the assumption that R is a tree. □

Because of this property it is easy to write algorithms that visit every node of a tree exactly once (see, for example, [Knuth 68, Section 2.3.1]). One could use the same algorithms on a general directed graph if one could identify a tree "hidden" in the directed graph (see Figure 1.4.3). Such a hidden tree is called a *spanning tree* of the graph.

Definition. If $G \subseteq A \times A$ is a directed graph and $R \subseteq A \times A$ is a tree and $R \subseteq G$, then we say R is a *spanning tree* of G.

As Figure 1.4.3 shows, a single graph may have many different spanning trees. If r is the root of a spanning tree R of G, there must be a path in R from r to x for every node in the tree. Since $R \subseteq G$, there must be a path in G from r to x for every node in G. Theorem 1.4.2 states that this is all that is needed for G to have a spanning tree.

FIGURE 1.4.3 A graph and some of its spanning trees.

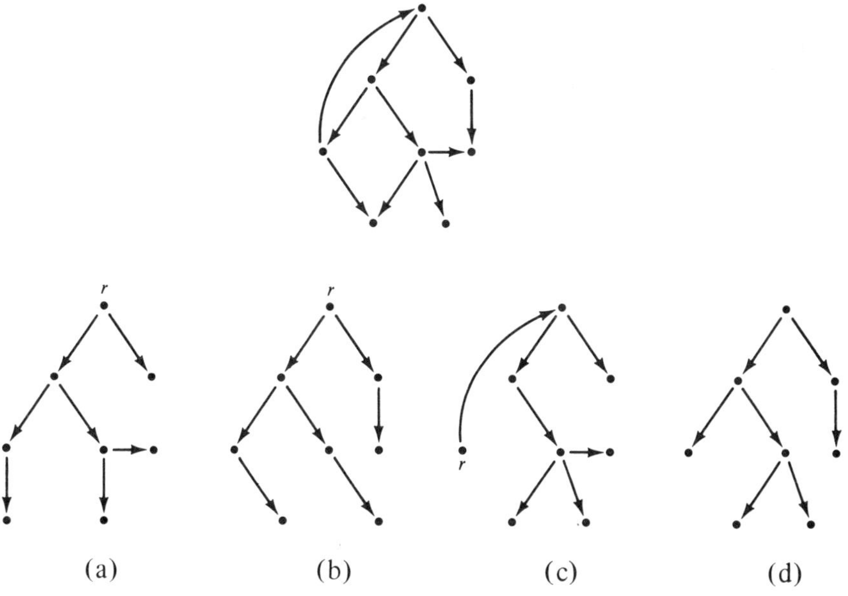

Definition. [16] $G \subseteq A \times A$ is a *rooted graph* iff there is a node $r \in A$ (the *root*) such that for every $x \in A$ there is a path from r to x in G.

Theorem 1.4.2. *Let $G \subseteq A \times A$ be a finite rooted graph with root r. Then G has a spanning tree with root r.*

PROOF. We shall construct a sequence of trees $R_k \subseteq A_k \times A_k$, each with root r and with $R_k \subseteq G$. Each A_k will contain k nodes, so if G has N nodes, $A_N = A$ and R_N will be a spanning tree for G.

First, let $A_1 = \{r\}$ and $R_1 = \varnothing$. (You can check that $R_1 \subseteq A_1 \times A_1$ is a tree.) Now imagine we have built $R_k \subseteq A_k \times A_k$, with $k < N$, and let us construct $R_{k+1} \subseteq A_{k+1} \times A_{k+1}$. Since A_k has k elements, and $k < N$, there must be some $z \in A$ such that $z \notin A_k$. Since G is rooted, let $\langle a_0, a_1, \ldots, a_p \rangle$ be a path in G from the root $r = a_0$ to $z = a_p$. Since $r \in A_k$, and $z \notin A_k$, there must be some j such that $a_0, a_1, \ldots, a_j$ all belong to A_k, but $a_{j+1} \notin A_k$.

Set $A_{k+1} = A_k \cup \{a_{j+1}\}$ and $R_{k+1} = R_k \cup \{(a_j, a_{j+1})\}$.

Now $a_{j+1} \notin A_k$, so (a_j, a_{j+1}) is the only arrow to a_{j+1}. So a_{j+1} has in-degree 1, and we could not have created a cycle by adding (a_j, a_{j+1}). Furthermore, $(a_j, a_{j+1}) \in G$, so $R_{k+1} \subseteq G$. Last, there is still no arrow ending at r, so r is the root of R_{k+1}. Hence R_{k+1} has the required properties. Perform this construction N times, and R_N will be the desired spanning tree. □

Figure 1.4.4 shows the construction of a spanning tree as described in the theorem. By selecting different z's and different paths, different spanning trees would have been constructed.

EXERCISES 1.4

1. Draw all the nonisomorphic trees with five vertices (see Exercise 1.3.6).

2. Let $R = \{(a,b)\}$, $A = \{a,b\}$, $B = \{a,b,c\}$. Is $R \subseteq A \times A$ a tree? Is $R \subseteq B \times B$ a tree?

3. Draw all the spanning trees of the following directed graph:

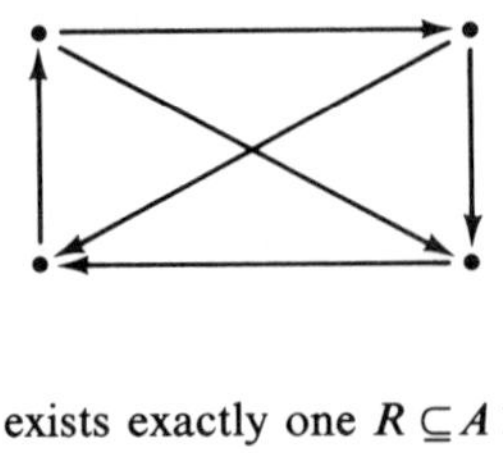

4. *Prove:* If $A = \{a\}$, there exists exactly one $R \subseteq A \times A$ that is a tree.

[16] This definition is also found in [Knuth 68, Section 2.3.4.2].

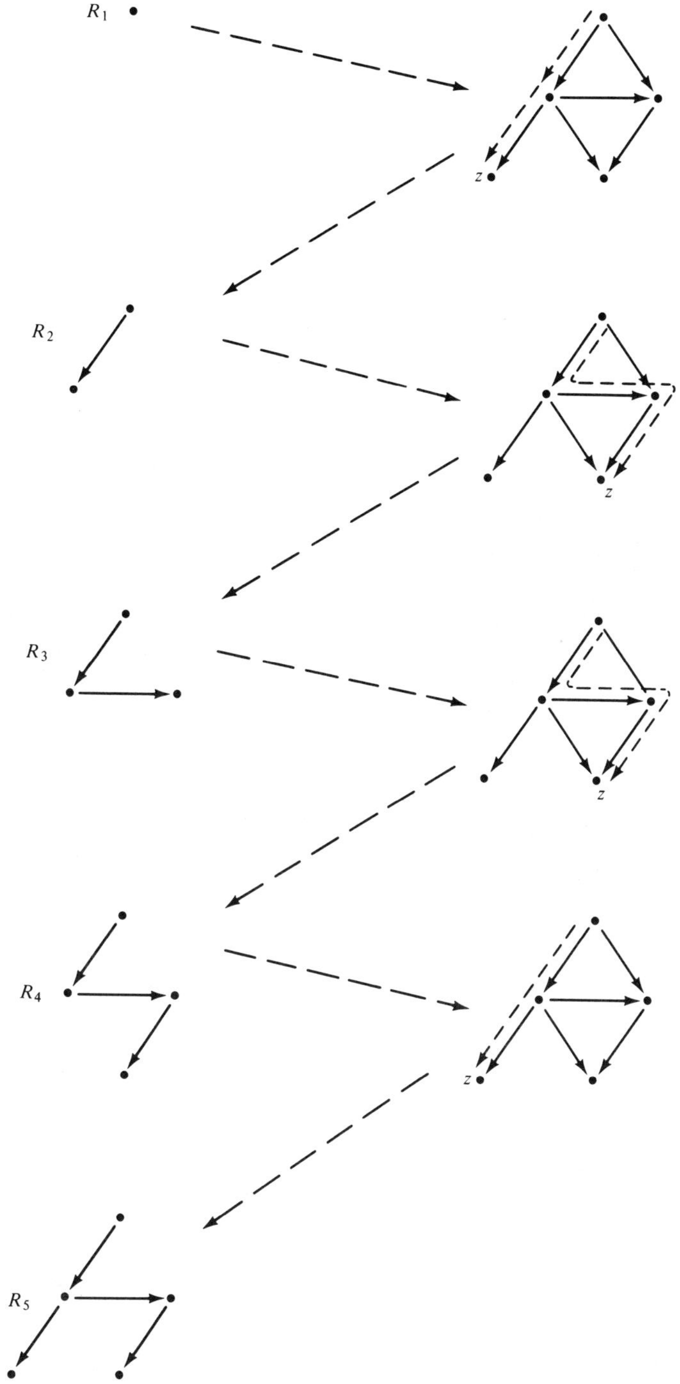

FIGURE 1.4.4 Construction of a spanning tree.

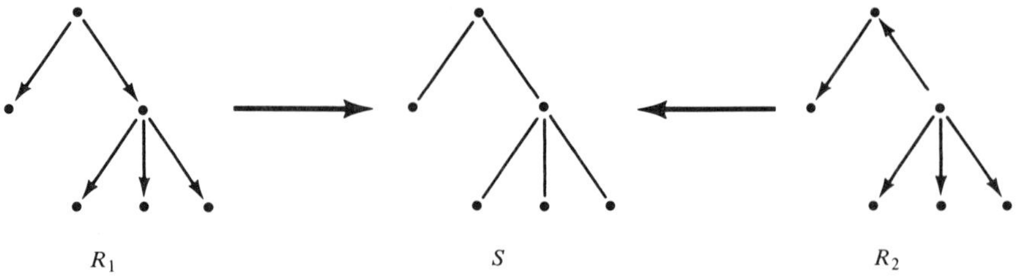

FIGURE 1.4.5 Two directed trees, R_1 and R_2, which yield the same undirected tree S.

1.5 REPRESENTATIONS OF GRAPHS

In this section we shall discuss several methods for representing directed graphs on a computer. In the process we shall get a brief preview of some fundamental data structuring methods.

It is meaningless to think about a representation of an object without considering how that representation is to be used. There are two basic kinds of operations one can perform on a piece of data (whether it be simple or complex)—one can change it (e.g., $x \leftarrow x+1$)[17] or one can test it for some property (e.g., $x>0$). To discuss our graph representations, we shall take two sample operations: adding an edge to the graph (a change), and testing for transitivity (a test).

We shall assume that we are interested in representing graphs with up to *maxnodes* nodes and *maxedges* edges. Furthermore, we shall assume that the nodes are numbered $1,\ldots,$ *numnodes*.

[17]We shall use a simple programming language in which "←" denotes assignment; complex statements are built up via

```
        begin S1;...;Sn end
        if B then S1 else S2
and     while B do S
```

Flow charts for these constructs are shown in Figure 6.1.1.

```
for i←a to b do S
```

is an abbreviation for

```
begin
    i←a;
    while i<b do begin S;
                       i←i+1
                 end
end
```

This is like a FORTRAN DO-loop except the loop body S may be executed zero times. This language is used again in Chapter 6.

Since a directed graph is just a relation $R \subseteq A \times A$, where A is the set of nodes and R is the set of edges, the simplest representation of a directed graph would be a 2-dimensional array *ADJ* with

$$ADJ[i,j] = \begin{cases} 1 & \text{if} \quad (i,j) \in R \\ 0 & \text{if} \quad (i,j) \notin R \end{cases}$$

This is called the *adjacency matrix* representation. For some applications this representation is quite convenient. To add an edge from node i to node j, we need only write

$$ADJ[i,j] \leftarrow 1$$

To test for transitivity, we could write:

```
trans←true;
for i←1 to numnodes do
    for j←1 to numnodes do
        for k←1 to numnodes do
            if ADJ[i,j]=1 & ADJ[j,k]=1 & ADJ[i,k]=0
              then trans←false
              else trans←trans
```

While this program is simple, it is spectacularly inefficient. It runs in time proportional to $(numnodes)^3$, regardless of the actual values in *ADJ*. We can improve it somewhat by replacing the assignment *trans*←false by a go-to[18] and by moving the $ADJ[i,j]=1$ test outside the innermost loop, but these are minor improvements. If the graph does not have many edges, the program is going to waste its time looking through the adjacency matrix trying to find them.[19] In this case, most of the entries in the adjacency matrix are 0, so we should suspect that there might be a more compact representation.

If there are few edges, we might do better to keep track of the edges directly. We therefore might number the arrows as well and create two 1-dimensional arrays *FROM* and *TO* (with size *maxedges*) by the following rule:

if the ith edge is (x,y), then $FROM[i]=x$ and $TO[i]=y$.

The adjacency matrix and *FROM–TO* table representations of a typical directed graph are shown in Figure 1.5.1.

This is surely more compact if $numedges < (numnodes)^2$, but it does not immediately help solve the problem of finding all the arrows coming out of

[18]Except that go-to's make a program harder to read and understand. Our little language does not have them. See [Dijkstra 68] and [Wirth 74] (for a dissenting view, read [Knuth 74]).

[19]Try to recall what you did when you hand-checked graphs for transitivity. Most likely you looked for edges, not nodes.

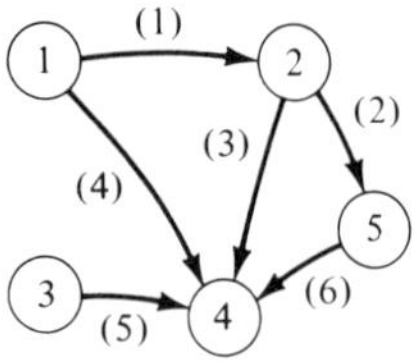

(a)

	1	2	3	4	5
1	0	1	0	1	0
2	0	0	0	1	1
3	0	0	0	1	0
4	0	0	0	0	0
5	0	0	0	1	0

(b)

	FROM	*TO*
1	1	2
2	2	5
3	2	4
4	1	4
5	3	4
6	5	4

(c)

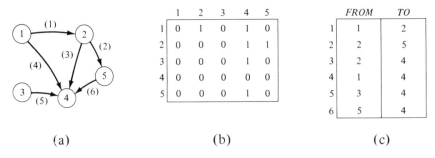

FIGURE 1.5.1 (a) A directed graph; (b) its adjacency matrix; (c) *FROM–TO* table.

a node. We can solve this problem by introducing another "parallel" array *NEXT* such that *NEXT*[*i*] contains the number of the next edge after *i* whose from-node is the same as edge *i* (or *NEXT*[*i*]=0 if there is no such edge). While we are at it, we shall introduce an array *NODES* such that *NODES*[*x*] contains the number of the first edge emanating from node *x*. This is called the *adjacency-list* representation, and is shown in Figure 1.5.2.

In order to find the out-degree of node *x*, we could write:

```
begin
    deg←0; p←NODES[x];
    while p≠0 do {deg is number of edges seen, not including p}
        begin {count p and advance}
            deg←deg+1;
            p←NEXT[p]
        end
end
```

To find whether there is an arrow from *x* to *z*:

```
begin
    p←NODES[x];
    if p=0 then found←false
        else begin
                while (TO[p]≠z)&(NEXT[p]≠0) do {p≠0}
                        p←NEXT[p];
                if TO[p]=z then found←true
                           else found←false
             end
end
```

We abbreviate this code by "*found*←*arrow*? (*x*,*z*)." To determine transitivity, we may now write the following code, which mimics the human

	NODES
1	1
2	2
3	5
4	0
5	6

	FROM	*TO*	*NEXT*
1	1	2	4
2	2	5	3
3	2	4	0
4	1	4	0
5	3	4	0
6	5	4	0

FIGURE 1.5.2 Adjacency-list representation of the directed graph in Figure 1.5.1(a).

transitivity algorithm of edge-tracing:

```
begin trans←true; {trans=true iff no violations have been found}
    for x←1 to numnodes do
        begin {look through edges out of x}
            e1←NODES[x];
            while (e1≠0) & trans do
                begin {look through edges out of e1's end}
                    y←TO[e1]; e2←NODES[y];
                    while (e2≠0) & trans do
                        begin
                                z←TO[e2];
                                found←arrow? (x,z);
                                if found then e2←NEXT[e2]
                                         else trans←false
                        end;
                    e1←NEXT[e1]
                end
        end
end
```

This code, while perhaps more complicated to read, does avoid the gross inefficiencies of the adjacency-matrix version.[20] Notice also that none of the code we have written uses the fact that $NEXT[e]$ is the *numerically* next edge whose from-node is the same as that of e. All that is necessary is that if we start with $e = NODES[i]$ and repeatedly perform $e \leftarrow NEXT[e]$ until e becomes 0, we shall get all the edges emanating from node i. Thus what we are really interested in are "slices" across the arrays, such as

FROM	*TO*	*NEXT*

[20]Notice how including *trans* in the while-loop tests causes quick termination when a violation is found.

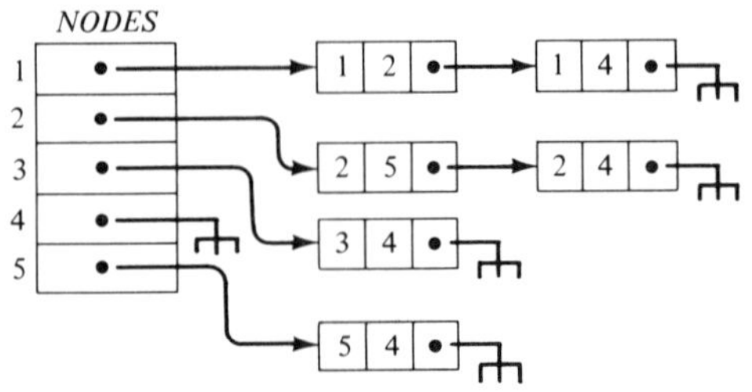

FIGURE 1.5.3 A better picture of the adjacency-list representation.

and we don't care how those slices are assembled into an array, so long as the *NEXT*-values are right. This leads us to the picture of Figure 1.5.3, which shows the adjacency-list representation of Figure 1.5.2 in terms of slices. Figure 1.5.4 shows a different array for the same adjacency list. This use of an extra field to link up array entries into lists is called the *linked-list* technique and is probably the most important single idea in the field of data structures.

Now we can finally write a program to add an edge from x to y. This operation increases the number of edges in the graph, which we assume is maintained in the variable *numedges*.

```
begin
        numedges←numedges + 1;
        FROM[numedges]←x;
        TO[numedges]←y;
        NEXT[numedges]←NODES[x];
        NODES[x]←numedges
end
```

A before-and-after diagram for this algorithm is shown in Figure 1.5.5.

Notice how the manipulations on these data structure are visualized in terms of graphs—not the graph in Figure 1.5.1(a), but the graph consisting of the nodes and edges of the data structure itself. The picture of a linked

FIGURE 1.5.4 Another way of assembling Figure 1.5.3 into an array.

	NODES
1	4
2	2
3	3
4	0
5	5

	FROM	*TO*	*NEXT*
1	1	4	0
2	2	5	6
3	3	4	0
4	1	2	1
5	5	4	0
6	2	4	0

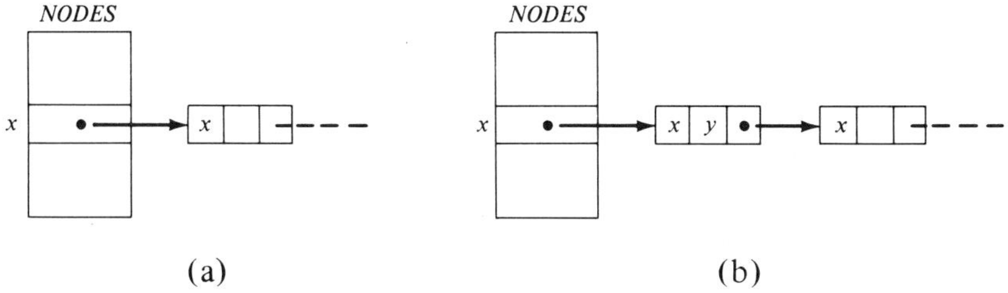

FIGURE 1.5.5 Inserting an edge from x to y, where (a) is before and (b) is after.

list as a graph with nodes and edges forms the connection between graph theory and data structures.

EXERCISES 1.5

1. (a) For each of the graph representations discussed in this section, write a program to find the in-degree of a node.
   (b) How could the representation be altered to make this easier? For your altered representation write a program to find the in-degree and the out-degree of a node.
   (c) For your altered representation, write a program to add an edge to the graph.

2. Consider a set of integers maintained as a linked list as follows:

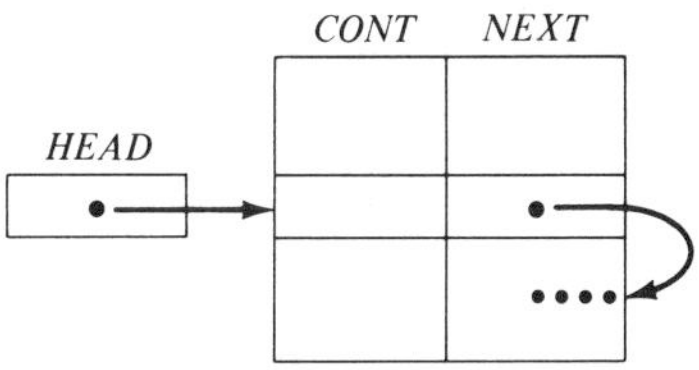

   Here, *HEAD* points at the first node in the list. Write programs to add an element to the set, delete an element, and test for membership in the set.

1.6 PERMUTATIONS AND COMBINATIONS

In this section we shall discuss some techniques for counting the number of elements of a set. This may sound like a trivial problem; but it is surely hard if the set has a large number of elements or if it is difficult to list the elements.

The number of elements in a set A is called its *cardinality*, and is denoted $|A|$. If $|A|$ is an integer, then we say A is *finite*; otherwise A is called *infinite*. Our discussion of element-counting techniques is limited to finite sets.[21] Rather than trying to present proofs, we shall list some examples of counting principles:

1. If A and B are disjoint finite sets, then $|A \cup B| = |A| + |B|$. For example, let

   $$A = \text{the set of lean people taking Comp. Sci. 101}$$
   $$B = \text{the set of mean people taking Comp. Sci. 101}$$

   If there are no people in CS101 who are both lean and mean, then the number of people who are either lean or mean is $|A| + |B|$.
2. If A and B are finite sets, then $|A \cup B| = |A| + |B| - |A \cap B|$. Our previous formula did not work because any elements of $A \cap B$ were counted twice—once as part of $|A|$ and once as part of $|B|$. By subtracting off $|A \cap B|$, every element gets counted exactly once. So, if in CS101 there are 31 lean people, 17 mean people, and four lean, mean people, the number of people who are either lean or mean is

   $$\begin{aligned} |A \cup B| &= |A| + |B| - |A \cap B| \\ &= 31 + 17 - 4 \\ &= 44 \end{aligned}$$
3. If A and B are finite sets, then $|A \times B| = |A| \times |B|$. We may justify this rule as follows: If $A = \{a_1, \ldots, a_m\}$ and $B = \{b_1, \ldots, b_n\}$, the elements of $A \times B$ are ordered pairs (a_i, b_j). We can arrange them in a rectangle according to the following diagram:

   Second component

		b_1		b_n
First component	a_1	(a_1, b_1)	$\cdots$	(a_1, b_n)
First component	a_2	(a_2, b_1)	$\cdots$	(a_2, b_n)
		$\vdots$		$\vdots$
First component	a_m	(a_m, b_1)	$\cdots$	(a_m, b_n)

   This rectangle has m rows and n columns; it has an area of $m \times n$.
4. If $A_1, \ldots, A_m$ are finite sets, then $|A_1 \times \cdots \times A_m| = |A_1| \times |A_2| \times \cdots \times |A_m|$. This is a straightforward generalization of the previous rule.
5. A *permutation* is a bijection from a set to itself. If $|A| = n$, then the number of permutations on A is $1 \times 2 \times \cdots \times (n-1) \times n = n!$.

   Let us try to justify this rule. Let $A = \{a_1, \ldots, a_n\}$, and let us try to define

[21] If A is infinite, there is still a notion of cardinality that makes sense, but it is considerably subtler than merely counting elements. For an elementary exposition of this theory, developed by Cantor in the 1890s, see [Stanat and McAllister 77, Chapter 6].

a permutation f by listing $f(a_1), f(a_2), \ldots, f(a_n)$. We may let $f(a_1)$ be any of the n elements of A; $f(a_2)$ may then be chosen from any of $n-1$ remaining elements. There are $n-2$ choices left for $f(a_3)$, and so on. There are therefore

$$n \times (n-1) \times (n-2) \times \cdots \times 2 \times 1$$

or $n!$ ways to permute the elements of A.[22]

A permutation is just a rearrangement of a set in a row.

6. Thinking of a permutation as an arrangement, we can generalize to the notion of a *permutation of n objects taken r at a time*, or a *selection of r objects out of n without repetition*. This corresponds to picking only $f(1)$ through $f(r)$, and for this there are

$$P(n,r) = n \times (n-1) \times \cdots \times (n-r+1) = \frac{n!}{(n-r)!}$$

choices, by the same argument as before.

7. A *combination of n objects taken r at a time* is an r-element subset of a set with cardinality n. Since we care only about selecting a subset, we do not care about the order of the elements in the subset. To illustrate the difference between a permutation and a combination, let $A = \{1,2,3,4,5\}$. Among the permutations of A taken 3 at a time are

124
142
214
241
412
421

but all of these correspond to the same three-element subset $\{1,2,4\}$. So $P(5,3)$ does not give the right number of combinations. Each combination is counted $3! = 6$ times, once for each possible way of listing the set. Therefore the number of combinations of five objects taken three at a time is

$$\frac{P(5,3)}{3!} = \frac{5 \times 4 \times 3}{3 \times 2} = 10$$

In general, to find the number of combinations of n objects taken r at a time, we can list the permutations of n objects r at a time, and then divide by $r!$, since each combination will be included $r!$ times in the listing. The number of combinations is therefore

$$\frac{P(n,r)}{r!} = \frac{n!}{r!(n-r)!}$$

[22] $n!$ is called the *factorial* of n.

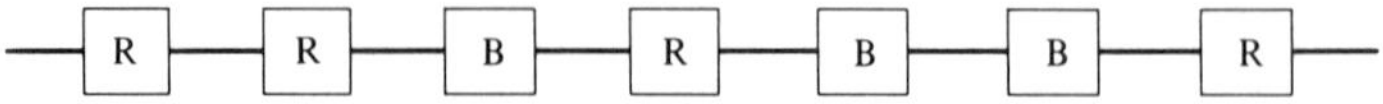

FIGURE 1.6.1

The symbol $\binom{n}{r}$ is used to denote this quantity. So the basic formula is

$$\binom{n}{r} = \frac{n!}{r!(n-r)!} = \frac{n \times (n-1) \times \cdots \times (n-r+1)}{r!}$$

The number of possible poker hands is

$$\binom{52}{5} = \frac{52 \times 51 \times 50 \times 49 \times 48}{4 \times 3 \times 2} = 2{,}598{,}960$$

8. The idea for the formula for $\binom{n}{r}$ was to find a set we knew how to count and then consider an equivalence relation on it. In that case, we took the selections of r out of n objects without repetition and said that two selections are equivalent iff one was a permutation of the other. There were $P(n,r)$ selections, and each equivalence class had $r!$ elements, so the number of equivalence classes was $P(n,r)/r!$. The same idea can be applied to other counting problems.

   For example, imagine we have four red beads and three blue beads, which we wish to string as in Figure 1.6.1.

   Assuming that beads of the same color are indistinguishable, and that we may not turn the string around or do any similar thing, how many ways can this be done?

   If the beads could be distinguished, there would be 7! ways of arranging them. Given such an arrangement, we may permute the red beads any of 3! ways and the blue beads any of 4! ways to get an equivalent arrangement. Hence the number of equivalence classes is

$$\frac{7!}{3!4!} = 35$$

For more on these ideas, see [Preparata and Yeh 73, Chapter 6] or [Berge 71].

EXERCISES 1.6

1. Imagine a programming language in which legal identifiers consist of a letter (a member of $\{\underline{A},\ldots,\underline{Z}\}$) followed by a digit (a member of $\{\underline{0},\ldots,\underline{9}\}$). How many distinct identifiers are there in the language?

2. Same as Exercise 1, except an identifier consists of a string of 1, 2, or 3 characters; the first character must be a letter and the others must be either letters or digits.

3. Calculate:

$$|\{n \mid 0 \leqslant n < 10^8 \text{ and the decimal representation of } n \text{ contains no 2's or 4's}\}|$$

4. In my CS101 class there are 25 lazy people, 60 hazy people, 15 lazy, hazy people, and 25 people who are neither lazy nor hazy. How many people are in the class?

5. In our department, every faculty member serves on exactly three committees. If there are seven committees to choose from, how many different committee assignments could a faculty member have?

6. In poker, a *full house* is a hand with two cards of one kind $(2,3,\ldots,10,\mathrm{J},\mathrm{Q},\mathrm{K},\mathrm{A})$ and three of another kind. For a conventional deck with four cards of each of the 13 kinds, how many different full houses are there?

7. An impoverished casino owner decides to install a roulette wheel with only 12 slots on the wheel. Six red slots, labeled with the odd numbers 1 through 11, alternate with six black slots, labeled with the even numbers 0 through 10. How many different such roulette wheels are there?

8. Write a computer program to generate all the permutations of the numbers $1,\ldots,n$.

9. Same as Exercise 8, except the permutations, read as n-ary numbers, should appear in numerical order.

10. Same as Exercise 8, except each permutation must differ from its predecessor in only two positions (i.e., by a transposition).

2 Induction and Its Applications

2.1 MATHEMATICAL INDUCTION

In previous sections, we did several constructions of the form: "Do the following until you are done" or "Imagine you have R_k; now build R_{k+1}, and so on."

Similarly, a program that searches an array X looks at $X[1]$, $X[2]$, "and so on." The precise mathematical equivalent of the phrase "and so on" is a principle called *mathematical induction*. It is perhaps the single most important idea in this book.

Principle. Let P be a predicate on ω. If

(i) $P(0)=\text{TRUE}$, and
(ii) for every k, if $P(k)=\text{TRUE}$, then $P(k+1)=\text{TRUE}$

then $P(n)=\text{TRUE}$ for all n.

Let us see why this is so. Assume (i) and (ii) hold. Then $P(0)=\text{TRUE}$. Since $P(0)=\text{TRUE}$, by (ii), then $P(1)=\text{TRUE}$. Since $P(1)=\text{TRUE}$, by (ii), then $P(2)=\text{TRUE}$. And so on.

So if we have to show $P(n)$ is always TRUE for some predicate P, we need only show (i) $P(0)=\text{TRUE}$ (the *base step*) and (ii) if $P(k)=\text{TRUE}$, then $P(k+1)=\text{TRUE}$ (the *induction step*). We call $P(k)$ the *induction hypothesis* (IH). Notice that in the induction step we do not assume $P(k)$ for *all* k. We show that if (by some good fortune) $P(k)=\text{TRUE}$ for some *particular* value of k, then $P(k+1)=\text{TRUE}$ as well.

EXAMPLE 1. For all $n\in\omega$,

$$0+1+\cdots+n=\frac{n(n+1)}{2}$$

PROOF. Let $P(k)=\text{TRUE}$ iff the proposition is true for $n=k$.

Base Step: $[P(0)=\text{TRUE}]$: $0=0\times 1/2=0$.

Induction Step: $[P(k)$ implies $P(k+1)]$: Assume $P(k)=\text{TRUE}$. Then

$$
\begin{aligned}
0+1+\cdots+k+(k+1) &= (0+\cdots+k)+(k+1) \\
&= \frac{k(k+1)}{2}+(k+1) \quad \text{(by IH)} \\
&= \frac{k(k+1)+2(k+1)}{2} \\
&= \frac{(k+1)(k+2)}{2}
\end{aligned}
$$

So $P(k+1)=\text{TRUE}$. □

EXAMPLE 2. There are infinitely many primes.

PROOF. Let $P(n)$ be "there are $>n$ primes."

Base Step: $P(0)=\text{TRUE}$ because there are >0 primes (in particular, 2 is a prime).

Induction Step: Assume $P(k)$, that is, there are $\geqslant k+1$ primes. Let $p_1,\ldots,p_{k+1}$ be $k+1$ primes. Let $N=(p_1\times\cdots\times p_{k+1})+1$. N is not divisible by any of the $k+1$ primes p_i. So either N is prime or else it has a prime divisor other than $p_1,\ldots,p_{k+1}$. So there are $>k+1$ primes. □

Let us redo one of our previous proofs and make it rigorous by using induction:

Proposition. If R is a transitive relation on A and there is a path from x to y in R, then $(x,y)\in R$.

PROOF. The induction hypothesis is, "For any $x,y\in A$, if there is a path of length k from x to y in R, then $(x,y)\in R$."

Base Step: $(k=1)$: A path of length 1 must be $\langle x,y\rangle$. By definition of a path, $(x,y)\in R$.

Induction Step: Assume the induction hypothesis is true for k. Take any path $\langle a_0,\ldots,a_k,a_{k+1}\rangle$ in R. Then $\langle a_0,\ldots,a_k\rangle$ is a path from a_0 to a_k of length k. By the induction hypothesis, $(a_0,a_k)\in R$. By the definition of path, $(a_k,a_{k+1})\in R$. So by transitivity $(a_0,a_{k+1})\in R$. □

Because programs contain loops, which are the programming equivalent of "and so on," induction is an important technique for proving assertions about programs.

Consider the following program:

```
begin
    x←x₀; y←y₀; z←0;
    {here is the top of the loop}
    while y≠0 do
        begin
            z←z+x;
            y←y−1
        end;
    ans←z
end
```

The flowchart for this program is shown in Figure 2.1.1. It multiplies x_0 times y_0 by adding the value of x_0 to z exactly y_0 times. (This only works, of course, if y_0 is a nonnegative integer). To prove this, we have to take account of the changing value of z, which we do in the following theorem:

Theorem 2.1.1. *In the program of Figure* 2.1.1, *if the program is started with* $y_0 \in \omega$, *then the program halts with* $ans = x_0 \times y_0$.

PROOF. The induction hypothesis is

IH(k): If for any values of x, y, and z, if the program ever reaches the top of the loop with $y \in \omega$ and $y = k$, then the program halts with $ans = z + x \times y$.

Before doing the induction, let us see how this will help us prove the theorem. If we do the induction, we can conclude:

P: For any values of x, y, and z, if the program ever reaches the top of the loop with $y \in \omega$, then the program halts with $ans = z + x \times y$.

(Notice that the $y = k$ is gone!) If we start the program with $y_0 \in \omega$, it immediately reaches the top of the loop with $x = x_0$, $y = y_0$, and $z = 0$. So, by assertion P, the program halts with

$$\begin{aligned} ans &= z + x \times y \\ &= 0 + x_0 \times y_0 \\ &= x_0 \times y_0 \end{aligned}$$

which is just what we wanted to prove.

Our sole remaining task is to prove assertion P by induction.

Base Step: ($k = 0$): If $y = 0$, then the program exits from the loop and sets $ans = z = z + x \times 0 = z + x \times y$.

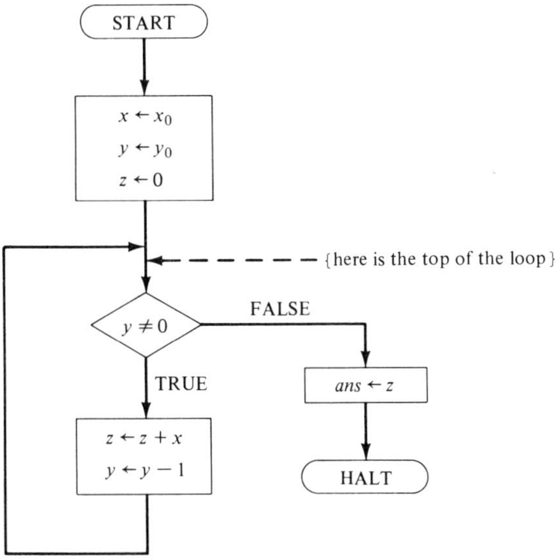

FIGURE 2.1.1

Induction Step: Assume IH(k). If the program reaches the top of the loop with $y=k+1$, then it executes the loop body and goes back to the top of the loop with new values in the three variables, given by

$$x'=x \quad \text{(unchanged)}$$
$$y'=y-1=k$$
$$z'=z+x$$

But now we have reached the top of the loop with the value of y equal to k, so IH(k) applies (!). So, by IH(k), the program eventually halts with $ans=z'+x'\times y'$. But

$$\begin{aligned} ans &= z'+x'\times y' \\ &= z+x+(x\times(y-1)) \\ &= z+x\times y \end{aligned}$$

so *ans* gets the right answer even when $y=k+1$. So IH($k+1$) is established, completing the induction step, and the proof.[1] □

[1]Notice the confusion in this proof between the *name* of a variable and the *value* it contains. This confusion is caused in part by our lack (at this stage of the book) of a good mathematical model of how programs really work. Such a model would account, among other things, for the name/value distinction. In Chapter 3 we shall develop such a model.

Mathematical induction is the tool that we will use most often. So stop now and make sure you understand it. An excellent little book, with many examples, is [Yousse 64]. [Knuth 68, Section 1.2.1] is another good discussion, with emphasis on the "algorithmic" aspect of induction. Most textbooks on "discrete structures," such as [Preparata and Yeh 73], [Stanat and McAllister 77], and [Prather 76] also contain sections on induction. Work several examples until you feel comfortable with this method of proof.

EXERCISES 2.1

Prove the following by induction.

1. $0+2+4+\cdots+2n=n(n+1)$.

2. $1+3+5+\cdots+2n+1=(n+1)^2$.

3. $\dfrac{1}{1\cdot 2}+\dfrac{1}{2\cdot 3}+\cdots+\dfrac{1}{(n+1)(n+2)}=\dfrac{n+1}{n+2}$.

4. $a+(a+d)+(a+2d)+\cdots+(a+nd)=\frac{1}{2}(n+1)(2a+nd)$.

5. $a+ar+ar^2+\cdots+ar^n=\dfrac{a(1-r^{n+1})}{1-r}$.

6. For any n, there exists $p\in\omega$ such that $2^p>n$.

7. What is wrong with the following proof?

Theorem. *For all nonnegative integers n, $a^n=1$.*

PROOF. Let IH(k) be, "if $n\leqslant k$, then $a^n=1$."

Base Step: $(k=0)$: $a^0=1$.

Induction Step: Assume IH(k). Then

$$a^{k+1}=\frac{a^k\times a^{k-1}}{a^{k-2}}=\frac{1\times 1}{1}=1$$

□

8. Prove that if the following program is started with $x_0\in\omega$, it halts with $ans=x_0+y_0$.

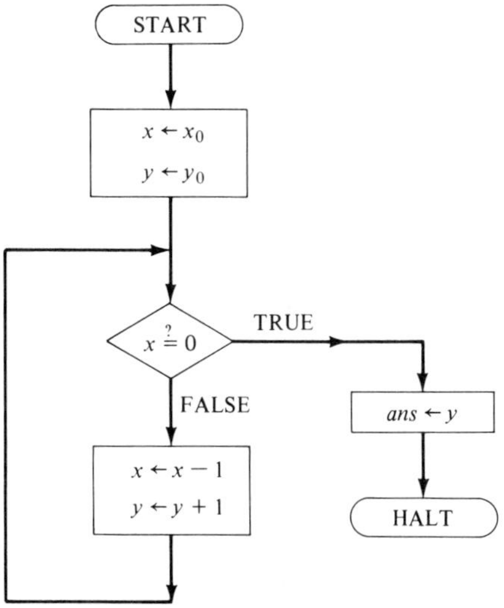

9.* (Loop Invariants) Consider the following flowchart, where B is some test (like $x \stackrel{?}{=} 0$) and S is some piece of program. Let P be a property (such as $x+y=x_0+y_0$) such that if P and B are true before S is executed, then P is still true afterward. Prove that if P is true at point a, and if the program ever reaches point c, then P is still true and B is false. (*Hint:* Use induction on the number of times S is executed to show that P is true every time the program reaches point a.)

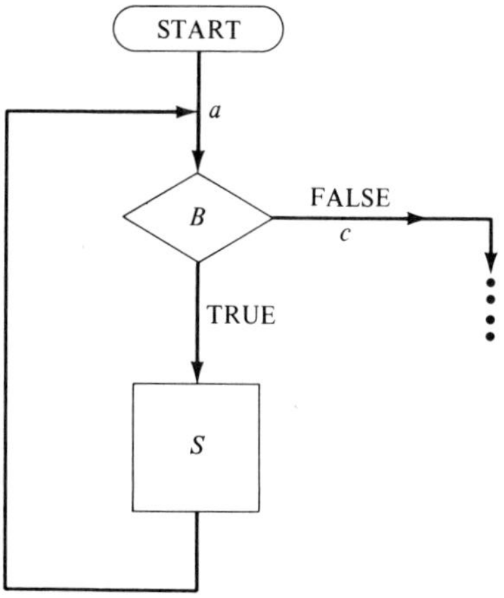

*Throughout, the asterisk indicates a more difficult exercise.

10. Letting P be the property $x+y=x_0+y_0$, use the result of Exercise 9 to prove that if the program of Exercise 8 halts, it does so with $ans=x_0+y_0$. (*Hint:* Use Exercise 9 to avoid doing a new induction.)

2.2 DEFINING SETS BY STAGES

The sets of interest in computer science tend to be rather complicated: consider $\{w\in V^+|w$ is a legal FORTRAN program$\}$. One of the things we shall be concerned with in this course is how to build complicated sets from simpler ones. We have already seen some set-combining methods in Section 1.1. These included product, union, intersection, and power sets.

As we defined them, product, union, and intersection were *binary* operations, that is, they combined two sets to produce a new set. There is, however, no good reason to restrict them to two arguments. So one can talk about the union of n sets.

$$A_1\cup\cdots\cup A_n=\{x\,|\,\text{for some } i\in\{1,\ldots,n\}, x\in A_i\}.$$

More generally, if I is any set, and for each $i\in I$ we have a set A_i, we can construct the union of the A_i, denoted $\bigcup_{i\in I} A_i$ or $\bigcup\{a_i|i\in I\}$, as

$$\{x\,|\,\text{for some } i\in I,\ x\in A_i\}$$

For example, if V_n denotes the set of all strings of length n over symbols in V, then $V^+=\bigcup_{n\in N} V_n$. Note that for any $j\in I$, $A_j\subseteq\bigcup_{i\in I} A_i$, and if $A_j\subseteq B$ for all j, then $(\bigcup_{i\in I} A_i)\subseteq B$.

One of the most common ways of constructing a complicated set is as a union of sets A_i where $i\in\omega$. We start off with a simple set A_0 and build up successively more complex sets A_k. Then we take the union of all the A_i (see Figure 2.2.1).

Let us try to construct a set of FORTRAN arithmetic expressions in this way. Let $A_0=\{w\in V^+|w$ is a legal FORTRAN constant or variable$\}$. A_0 is the simplest set of arithmetic expressions. Call these "complexity zero." Now let us use A_0 to build a set of slightly more complicated expressions. Say an expression of complexity 1 is an expression of complexity 0 or the sum of two expressions of complexity 0:

$$A_1=A_0\cup\{w\underline{+}v\,|\,w,v\in A_0\}$$

So A_1 consists of anything that is a constant, a variable, or the sum of variables and constants. Now say an expression of complexity 2 is an expression of complexity 1 or the sum of two expressions of complexity 1:

$$A_2=A_1\cup\{w\underline{+}v\,|\,w,v\in A_1\}$$

Now we can build A_3 from A_2, A_4 from A_3, and so on, building A_{k+1} from A_k:

$$A_{k+1}=A_k\cup\{w\underline{+}v\,|\,w,v\in A_k\}$$

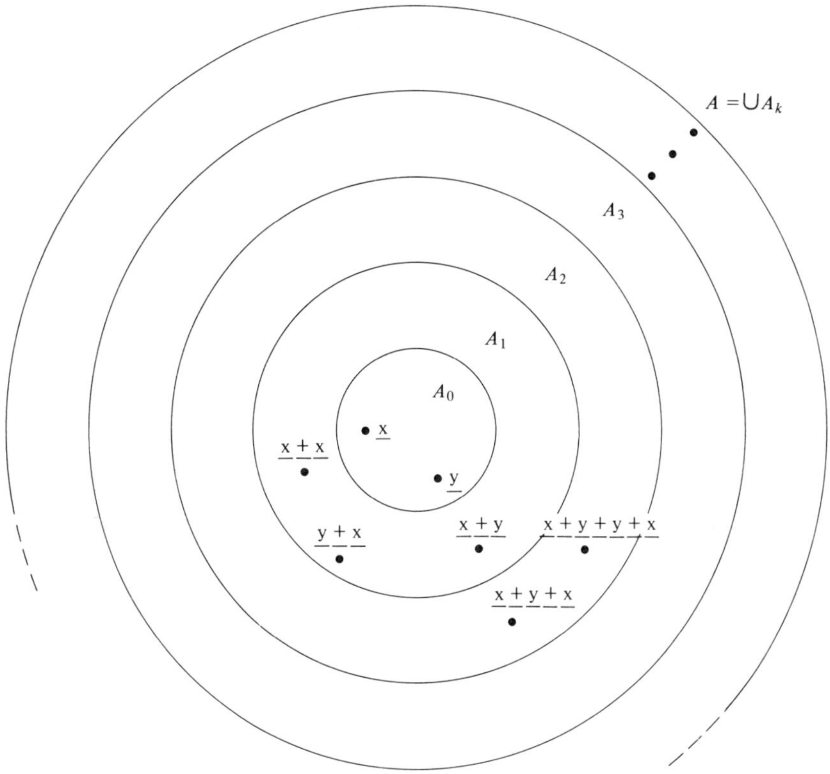

FIGURE 2.2.1

Now we can take $\bigcup_{k \in \omega} A_k$. This is just the set of all expressions built up from constants and variables by addition.[2]

Let us try one more example.

Let $A_0 = \{(0,0)\}$. Let $A_{k+1} = A_k \cup \{(x+1, y+2) | (x,y) \in A_k\}$. Then

$$A_1 = A_0 \cup \{(x+1, y+2) | (x,y) \in A_0\}$$
$$= \{(0,0)\} \cup \{(1,2)\} = \{(0,0),(1,2)\}$$
$$A_2 = A_1 \cup \{(x+1, y+2) | (x,y) \in A_1\}$$
$$= \{(0,0),(1,2)\} \cup \{(x+1, y+2) | (x,y) \in \{(0,0),(1,2)\}\}$$
$$= \{(0,0),(1,2)\} \cup \{(1,2),(2,4)\}$$
$$= \{(0,0),(1,2),(2,4)\}$$

[2]Of course, we have not captured *all* FORTRAN expressions. We could have gotten them all if we had been sufficiently careful in building A_{k+1} from A_k.

$$A_3 = \{(0,0),(1,2),(2,4),(3,6)\} \quad \text{(similarly)}$$
$$A_4 = \{(0,0),(1,2),(2,4),(3,6),(4,8)\}$$

So $\bigcup_k A_k = \{(x,2x)|x \in \omega\}$.

In this case we can get an explicit definition of A_k:

$$A_k = \{(x,2x)|x \in \omega \text{ and } x \leqslant k\}$$

Each A_k is a "portion" of $\bigcup_k A_k$ that is "bounded" by k. To get $\bigcup_k A_k$ we simply remove the bound.[3] Notice also that the notation $\bigcup_k A_k$ is in effect a subroutine with k as its local variable; therefore, k cannot appear in $\{(x,2x)|x \in \omega\}$, which is the result of the subroutine.[4]

This is a relation on ω. In fact, it happens to be a function. So not only can we build sets in stages, but we can define relations and functions in the same way. We will see more of this technique in Section 4.

EXERCISES 2.2

For each of the following, describe $\bigcup_{k \in \omega} A_k$.

1. $A_k = \{0,\ldots,k\}$.
2. $A_k = \{2j|j \in \omega \& 0 \leqslant j \leqslant k\}$.
3. $A_k = \{j|j \in Z \& -k \leqslant j \leqslant k\}$.
4. $A_k = \{p|p \text{ is a prime number} \leqslant k\}$.

2.3 DEFINING SETS BY INDUCTION

As good mathematicians, we now analyze the procedure we have just used by attempting to simplify and remove extraneous details. In both cases we had a set U (V^+ and $\omega \times \omega$, respectively). We started off with a set $A_0 \subseteq U$ and defined A_{k+1} to be $A_k \cup \{f(x)|x \in A_k\}$ for some f: $U \to U$. The final result was $A = \bigcup_{k \in \omega} A_k$. A is "closed under f" in the following sense:

Definition. If f: $U \to U$, and $S \subseteq U$, we say S is *closed under f* iff

$$\text{if} \quad x \in S \quad \text{then} \quad f(x) \in S$$

In general, given a set U and f: $U \to U$, there will be many subsets of U closed under f. The empty set is always closed under f, as is all of U. Of course, the empty set is not very interesting, since it does not contain

[3] This is what we did in the proof of Theorem 2.1.1.

[4] Similarly, if you are asked to evaluate some definite integral $\int_a^b f(x)dx$, you would not expect the variable x to appear in the answer.

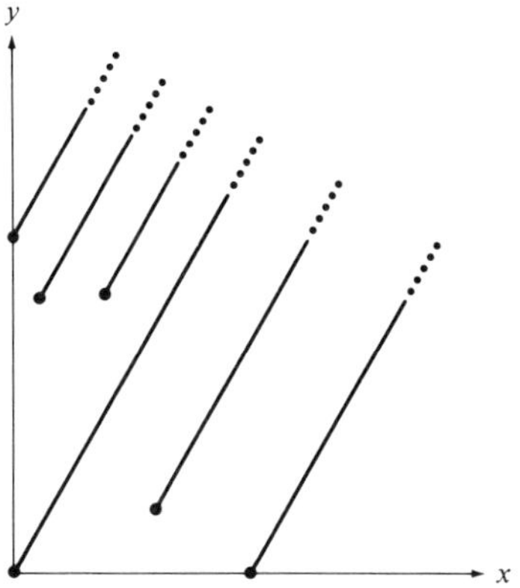

FIGURE 2.3.1 A typical subset of $\omega\times\omega$ closed under f and containing $A_0=\{(0,0)\}$. Note that only the integer-valued points are included. The lines have a slope of 2.

anything. So we might do better to look at those subsets of U that are closed under f and that contain A_0 as a subset.

Even so, there are usually lots of subsets S with $A_0 \subseteq S$ and S closed under f. For example, let

$$U=\omega\times\omega$$
$$f=\{(x+1,y+2)|(x,y)\in\omega\times\omega\}$$
$$A_0=\{(0,0)\}$$

as in the last section. Then each of the following subsets of $\omega\times\omega$ is closed under f and has A_0 as a subset:

(i) $\{(x,2x)|x\in\omega\}$
(ii) $\{(x,2x)|x\in\omega\}\cup\{(x,2x+5)|x\in\omega \text{ and } x\geqslant 3\}$
(iii) $\{(x,2x)|x\in\omega\}\cup\{(x,2x+3)|x\in\omega \text{ and } x\geqslant 2\}$
$\cup\{(x,2x-5)|x\in\omega \text{ and } x\geqslant 3\}$

A "typical" subset of $\omega\times\omega$ closed under f and containing A_0 is shown in Figure 2.3.1. It consists of the integer points on lines with a slope of 2.

Among all these subsets of $\omega\times\omega$, the set A that we built by stages has a special property: it is a *subset* of each of these subsets. Loosely speaking, it is the "smallest" subset of U that contains A_0 as a subset and that is closed under f. We make this observation into a theorem:

Theorem 2.3.1 (Fundamental Theorem on Induction). *Let U be a set, $C\subseteq U$, f: $U\to U$. Let*

$$A_0=C$$
$$A_{k+1}=A_k\cup\{f(x)|x\in A_k\}$$
$$A=\bigcup\{A_k|k\in\omega\}$$

Then

(i) *A is closed under f*

(ii) *if S is any subset of U that has A_0 as a subset and is closed under f, then $A \subseteq S$*

PROOF. (i) To show A is closed under f, we must show that if $x \in A$, then $f(x) \in A$. If $x \in A$, then for some k, $x \in A_k$. But then $f(x) \in A_{k+1} \subseteq A$. So $f(x) \in A$.

(ii) Let S be a subset of U closed under f, with $A_0 \subseteq S$. We want to show that $A \subseteq S$. It will suffice to show that for each k, $A_k \subseteq S$, for then $\bigcup A_k = A \subseteq S$. We proceed by induction. Let $P(k)$ be the predicate "$A_k \subseteq S$."

Base Step: $A_0 \subseteq S$, by assumption.

Induction Step: Assume $A_k \subseteq S$. We must show that if $y \in A_{k+1}$, then $y \in S$. If $y \in A_{k+1}$, then either $y \in A_k$ or $y \in \{f(x) | x \in A_k\}$. If $y \in A_k$, then $y \in S$ by IH. If $y \in \{f(x) | x \in A_k\}$, then $y = f(x^*)$ for some $x^* \in A_k$. By IH, $x^* \in S$. By closure of S under f, $f(x^*) = y \in S$. □

A picture of the construction is shown in Figure 2.3.2.

Because of this theorem, we may now introduce a new notation for defining sets:

Definition Scheme. Let U be a set, $C \subseteq U$, f: $U \to U$. Then the following language

(i) $C \subseteq A$

(ii) if $x \in A$, then $f(x) \in A$

(iii) nothing else

defines $A = \bigcup \{A_k | k \in \omega\}$ as before.

That is to say, when we write (i)–(iii), we mean the smallest set satisfying (i)–(iii), which is just the A of the theorem.

So now we write our two examples as follows:

EXAMPLE 1

(i) {constants and variables} $\subseteq A$

(ii) if w and $v \in A$, then so is $w \pm v$

(iii) nothing else[5]

[5]Strictly speaking, Example 1 does not fit the form of our definition, since here the f is a function from $V^+ \times V^+$ to V^+, instead of from V^+ to V^+ as required by the theorem. The appropriate generalization of Theorem 2.3.1 is stated in Exercise 2.3.6. Similar remarks may

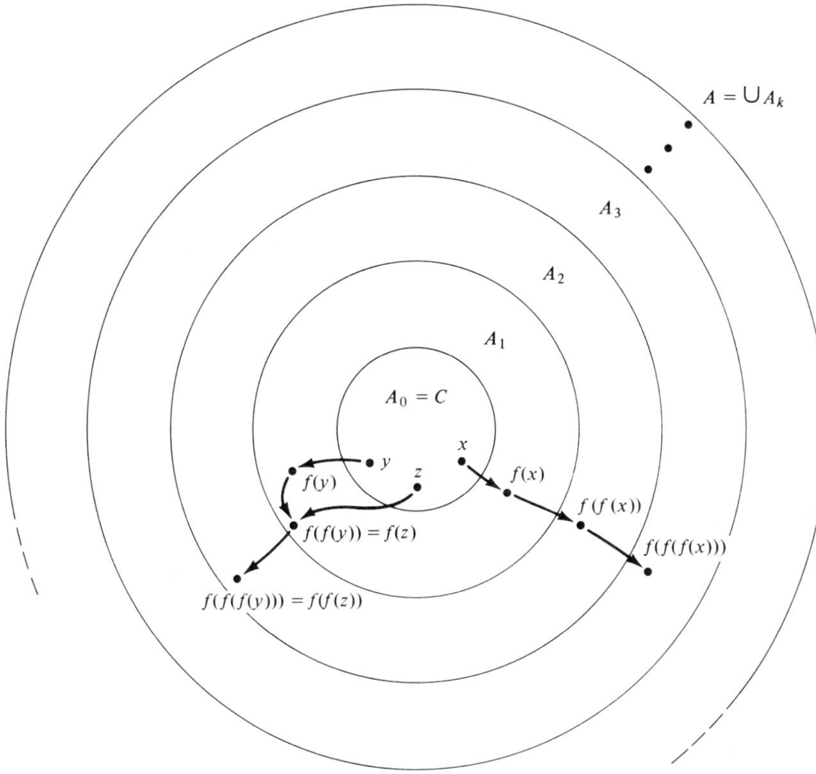

FIGURE 2.3.2

EXAMPLE 2

(i) $(0,0) \in A$
(ii) if $(x,y) \in A$, then $(x+1, y+2) \in A$
(iii) nothing else

Now let us use the theorem to show that the A of Example 2 is exactly $\{(x,2x) | x \in \omega\}$, which we claimed, but did not prove, at the end of Section 2.2.

Let $S = \{(x,2x) | x \in \omega\}$. We want to show $A = S$. We claim S is closed under f, where $f(x,y) = (x+1, y+2)$. To show the claim, we note that if $(x,2x) \in S$, then $f(x,2x) = (x+1, 2x+2) = (x+1, 2(x+1)) \in S$. Since $\{(0,0)\} \subseteq S$, by the theorem, $A \subseteq S$. We must now show $S \subseteq A$. We show,

---

be made about Exercise 2.3.4, which has several f's. Here the appropriate generalization is stated in Exercise 2.3.5 (set $C = D = U$). The remainder of the book will always assume an appropriate version of the fundamental theorem is available.

by induction, that $(k,2k)\in A$: $(0,0)\in A$; and if $(k,2k)\in A$, then, by (ii), $(k+1,2k+2)=(k+1,2(k+1))\in A$.

Moral: When a set is defined by induction, you can prove things about it by induction.

EXERCISES 2.3

Describe the following inductively defined subsets of ω:

1. $0\in A$
if $x\in A$, then $x+1\in A$
nothing else

2. $1\in A$
if $x\in A$, then $2x\in A$
nothing else

3. $0\in A$
if $x\in A$, then $2x\in A$
nothing else

4. $1\in A$
if $x\in A$, then $2x\in A$
if $x\in A$, then $3x\in A$
nothing else

5. Modify the proof of Theorem 2.3.1 to prove the following:

Let U be a set, $C,D\subseteq U$, f,g: $U\to U$. Let

$$A_0=C$$
$$A_{k+1}=A_k\cup\{f(x)|x\in A_k\cap D\}\cup\{g(x)|x\in A_k\cap D\}$$

Then $\bigcup_{k\in\omega}A_k$ is the smallest set S such that

(i) $C\subseteq S$
(ii) if $x\in S\cap D$, then $f(x)\in S$
(iii) if $x\in S\cap D$, then $g(x)\in S$

6. Modify the proof of Theorem 2.3.1 to prove the following:

Let U,V be sets, f: $U\times V\to U$, g: $U\times V\to V$, $C\subseteq U$, $D\subseteq V$. Let

$$A_0=C$$
$$B_0=D$$
$$A_{k+1}=A_k\cup\{f(x,y)|x\in A_k \,\&\, y\in B_k\}$$
$$B_{k+1}=B_k\cup\{g(x,y)|x\in A_k \,\&\, y\in B_k\}$$
$$A=\bigcup_{k\in\omega}A_k$$
$$B=\bigcup_{k\in\omega}B_k$$

Then A and B are the smallest sets S and T such that

(i) $C \subseteq S$
(ii) $D \subseteq T$
(iii) if $x \in S$ and $y \in T$, then $f(x,y) \in S$
(iv) if $x \in S$ and $y \in T$, then $g(x,y) \in T$

7. We say that a relation is a *preorder* iff it is reflexive and transitive. *Prove:* For any relation R on A, there exists a relation R^* on A with the following properties:

(i) R^* is a preorder on A, and $R \subseteq R^*$
(ii) If S is a preorder on A, and $R \subseteq S$, then $R^* \subseteq S$

8. Let U be a set, f: $U \rightarrow U$, $C \subseteq U$, and P: $U \rightarrow \{\text{TRUE}, \text{FALSE}\}$. Let A be defined by

(i) $C \subseteq A$
(ii) if $x \in A$, then $f(x) \in A$
(iii) nothing else

Prove: If for every $c \in C$, $P(c) = \text{TRUE}$, and if $P(x) = \text{TRUE}$, then $P(f(x)) = \text{TRUE}$, *then* for every $a \in A$, $P(a) = \text{TRUE}$.

9. State and prove the best generalization of Theorem 2.3.1 that you can think of.

2.4 DEFINING FUNCTIONS BY INDUCTION

When a set is defined by induction, the best way to define a function from it to another set is by induction.

Let us try to define a function M that, given an arithmetic expression, yields the value of the expression. We must first define the set A of expressions to be evaluated. Let $V = \omega \cup \{\underline{+}, \underline{(}, \underline{)}\}$. A will be a subset of V^+; that is, every expression will be a string whose characters are either integers[6] or one of $\underline{+}$, $\underline{(}$, or $\underline{)}$. Define A as follows:

(i) $\omega \subseteq A$
(ii) if $x,y \in A$, $\underline{(}x\underline{)}\underline{+}\underline{(}y\underline{)} \in A$
(iii) nothing else[7]

[6]Graduates of "the new math" should object at this point that one cannot write down an integer; one can only write down a *numeral* representing an integer. We could do so, but only at the expense of obscuring the main point by the additional mechanics. The treatment of numerals is sketched in Exercise 2.4.7.

[7]When the universe is V^+ and the building function f is some flavor of concatenation, the standard notation gives this for our example:

$$\langle \text{A} \rangle ::= \langle \text{nonnegative integer} \rangle$$
$$\langle \text{A} \rangle ::= \underline{(}\langle \text{A} \rangle\underline{)} \ \underline{+} \ \underline{(}\langle \text{A} \rangle\underline{)}$$

Here names of sets are surrounded by angle brackets $\langle \ldots \rangle$, the x and y have been eliminated, and the $::=$ symbol replaces "$\in$" (with left and right sides switched). This notation is called

Typical members of A are

$$3$$

$$\underline{(}5\underline{)}\ \underline{+}\ \underline{(}8\underline{)}$$

$$\underline{(}\ \underline{(}\ \underline{(}7\underline{)}\ \underline{+}\ \underline{(}3\underline{)}\ \underline{)}\ \underline{+}\ \underline{(}4\underline{)}\ \underline{)}\ \underline{+}\ \underline{(}3\underline{)}$$

Typical nonmembers of A are

$$-4$$

$$5\ \underline{+}\ \underline{(}7\underline{)}$$

M is to be a function $A \rightarrow \omega$, that is, a subset of $A \times \omega$. So we may define M as follows:

(i) $\{(k,k) | k \in \omega\} \subseteq M$
(ii) if (x,k) and (y,k') are members of M, then $(\underline{(}x\underline{)}\underline{+}\underline{(}y\underline{)}, k+k') \in M$
(iii) nothing else

Using the more familiar notation $M(x)=k$ for $(x,k) \in M$, we may paraphrase this definition as follows[8]:

(i) $M(k)=k$ for $k \in \omega$
(ii) $M(\underline{(}x\underline{)}\underline{+}\underline{(}y\underline{)}) = M(x)+M(y)$

We sometimes refer to the first definition as being in *relational form* because it emphasizes that M is just an inductively defined set of ordered pairs—that is, a relation. We call the second definition the *functional form* because it emphasizes the computation of M in terms of its arguments and values.

Backus–Naur form or BNF, after John Backus and Peter Naur, who used it to specify the syntax of ALGOL 60. For a concise discussion of BNF, see [Pratt 75, pp. 301–306]. BNF is a particular notation for a *context-free grammar*. For more on grammars, see [Denning, Dennis, and Qualitz 78]. We have not used BNF in this book because we shall need to have x and y in the definition.

[8]Why are some of these parentheses underscored and some not? We are defining a language (the set A) and a function on it (the function M). Among the symbols in this language are the parentheses $\underline{(}$ and $\underline{)}$; symbols in this language are always underscored. We are also writing down our definition in some language, in this case the language of mathematical English. The language in which the definition is written is called the *metalanguage*. Among the symbols in the metalanguage are the usual parentheses (and); symbols in the metalanguage are never underscored. Another feature of our metalanguage is that it includes variables, such as k, x, and y in this example, which range over strings in the defined language. Such variables are called *metalinguistic variables* [since they are variables which range over the strings in the language (cf. "real variable")] or just *metavariables*. We have used metavariables before: see Chapter 1, footnote 11.

Thus, the content of footnote 7 is that BNF is another metalanguage. Our metalanguage is informal in the sense that we shall not attempt to give a rigorous description of it; such a description would perforce be written in a metametalanguage (!).

To contrast the two forms, let us compute M on the string $((7)+((3)+(4)))+(3)$, using both forms. In the relational form, the computation strictly follows the inductive definition:

$$
\begin{array}{rl}
3 \in A & (3,3) \in M \\
4 \in A & (4,4) \in M \\
7 \in A & (7,7) \in M \\
(3)+(4) \in A & ((3)+(4),7) \in M \\
(7)+((3)+(4)) \in A & ((7)+((3)+(4)),14) \in M \\
((7)+((3)+(4)))+(3) \in A & (((7)+((3)+(4)))+(3),17) \in M
\end{array}
$$

So $M(((7)+((3)+(4)))+(3))=17$, which is as it should be. The functional form, on the other hand, consists of algebraic identities that describe the function M. To evaluate M on a string we apply whatever identities we can until an answer is reached.

By the construction of A, either $w \in \omega$ or else $w=(x)+(y)$ for some $x,y \in A$. If $w=k \in \omega$, then $M(w)=k$. If $w=(x)+(y)$, $M(w)$ is the sum (in ω) of $M(x)$ and $M(y)$. So to evaluate M of a long string in A, look for the "main connective," evaluate each side, and then add the results: $w=((7)+((3)+(4)))+(3)$ is in A. The "main connective" is the rightmost "$+$"; that is, $w=(x)+(y)$ where $x=(7)+((3)+(4))$ and $y=3$. So

$$
\begin{aligned}
&M(((7)+((3)+(4)))+(3)) \\
&=M((7)+((3)+(4)))+M(3) \\
&=(M(7)+M((3)+(4)))+M(3) \\
&=(M(7)+(M(3)+M(4)))+M(3) \\
&=(M(7)+(3+4))+M(3) \\
&=(7+7)+M(3) \\
&=14+3 \\
&=17
\end{aligned}
$$

The student should understand this example thoroughly, as we shall be defining functions by induction and calculating with them throughout this book. Let's do another example:

$E \subseteq \omega$	$g\colon E \to \omega$
(i) $0 \in E$	(i) $g(0)=0$
(ii) if $x \in E$, then $(x+2) \in E$	(ii) $g(x+2)=g(x)+1$
(iii) nothing else	

Clearly, E is the set of even numbers and $g(x)=x/2$, but let's follow a calculation in detail:

$$\begin{aligned} g(6) &= g(4)+1 \\ &= (g(2)+1)+1 \\ &= ((g(0)+1)+1)+1 \\ &= ((0+1)+1)+1 \\ &= 3 \end{aligned}$$

There are, of course, pitfalls in proceeding in this manner. Define a set of arithmetic expressions in $(\omega \cup \{\underline{+}, \underline{*}\})^+$ as follows:

$E \subseteq V^+$	$M: E \to \omega$
(i) $\omega \subseteq E$	(i) $M(k)=k$ for $k \in \omega$
if $x,y \in E$, then	
(ii) $x \underline{+} y \in E$	(ii) $M(x \underline{+} y)=M(x)+M(y)$
(iii) $x \underline{*} y \in E$	(iii) $M(x \underline{*} y)=M(x)\times M(y)$

This looks promising, but it fails:

$$M(2)=2, \qquad M(3)=3, \qquad M(5)=5$$

$$M(2 \underline{*} 3)=M(2)\times M(3)=2\times 3=6$$

$$M(3 \underline{+} 5)=M(3)+M(5)=3+5=8$$

$$M(2 \underline{*} 3 \underline{+} 5)=M(2 \underline{*} 3)+M(5)=6+5=11$$

$$M(2 \underline{*} 3 \underline{+} 5)=M(2)\times M(3 \underline{+} 5)=2\times 8=16$$

So $11=16$. (!?)

What is the problem here? The difficulty is that there are two different ways to create $2 \underline{*} 3 \underline{+} 5$. We say that such a string is *ambiguous*. We can often remedy the situation by changing the definition of E. The following rules can be seen to generate correct arithmetic expressions:

$F,T,E, \subseteq (\omega \cup \{\underline{(}, \underline{)}, \underline{+}, \underline{*}\})^+$	$M_F: F\to\omega,\ M_T: T\to\omega,\ M_E: E\to\omega$
(i) $\omega \subseteq F$	$M_F(k)=k$, for $k\in\omega$
(ii) if $t\in E$, then $\underline{(}\, t\, \underline{)} \in F$	$M_F(\underline{(}\, t\, \underline{)})=M_E(t)$
(iii) $F\subseteq T$	$M_T(t)=M_F(t)$ for $t\in F$
(iv) if $x\in T$ and $y \in F$, then	
$x \underline{*} y \in T$	$M_T(x \underline{*} y)=M_T(x)\times M_F(y)$
(v) $T\subseteq E$	$M_E(t)=M_T(t)$ for $t\in T$
(vi) if $x\in E$ and $y\in T$, then	
$x \underline{+} y \in E$	$M_E(x \underline{+} y)=M_E(x)+M_T(y)$

F, T, and E stand for factors, terms, and expressions, respectively. One can prove (by induction, of course) that every string in E is unambiguous. So let us do our pathological example:

$$\begin{aligned}M_E(2 \underline{*} 3 \underline{+} 5) &= M_E(2 \underline{*} 3) + M_T(5)\\ &= M_T(2 \underline{*} 3) + M_F(5)\\ &= (M_T(2) \times M_F(3)) + 5\\ &= (M_F(2) \times 3) + 5\\ &= (2 \times 3) + 5\end{aligned}$$

You may confirm that under these rules, not only does multiplication get done before addition, but consecutive additions or multiplications get performed from left to right, and calculations inside parentheses also get done first. You should become facile with these calculations as we shall be spending a great deal of time defining sets of strings (programs!) and functions on them (output!) by induction.

EXERCISES 2.4

1.* (a) Using the "functional" definition of M on page 46 calculate

$$M(\underline{(}\,\underline{(}4\underline{)}\,\underline{+}\,\underline{(}2\underline{)}\,\underline{)}\,\underline{+}\,\underline{(}\,\underline{(}5\underline{)}\,\underline{+}\,\underline{(}6\underline{)}\,\underline{)})$$

(b) Confirm, using the definition of A on page 45, that the expression is in A.
(c) Using the "relational" definition of M, repeat part (a).

Convert the following definitions of relations from $\omega \times \omega$ to ω into functional form:

2. (i) for all y, $((0,y),0) \in F$
(ii) if $((x,y),k) \in F$, then $((x+1,y),k+y) \in F$
(iii) nothing else

3. (i) for all x, $((x,0),x) \in G$
(ii) for all y, $((0,y),0) \in G$
(iii) if $((x,y),k) \in G$, then $((x+1,y+1),k) \in G$

4. For F and G defined above, compute in relational form

(a) $F(2,3)$
(b) $G(5,2)$
(c) $G(2,4)$

5. Repeat Exercise 4 using the functional forms from Exercises 2 and 3.

6. Consider the following functional definition:

(i) if $x > 100$, then $F(x) = x - 10$
(ii) if $x \leqslant 100$, then $F(x) = F(F(x+11))$

Compute $F(97)$ using functional form.

7. Consider the set $\text{Num} \subseteq \{\underline{0}, \underline{1}\}^+$ defined as follows:

   (i) $\underline{1} \in \text{Num}$
   (ii) if $x \in \text{Num}$, $x\underline{0} \in \text{Num}$
   (iii) if $x \in \text{Num}$, $x\underline{1} \in \text{Num}$

   Num is clearly the set of binary numerals with a leading $\underline{1}$. In either functional or relational form, write a function B: Num$\rightarrow\omega$ that sends each numeral to the integer it represents.

8. In your favorite programming language, write a program that reads a sequence of 0's and 1's and translates the sequence into an integer.

9. Let $f \subseteq \omega \times \omega$ be defined as follows:

   1. $(0,2) \in f$
   2. if $(x,y) \in f$, then $(x+1, y+3) \in f$
   3. nothing else

   Prove that $f = \{(x, 3x+2) | x \in \omega\}$.

10*. Consider the function F defined in Exercise 6. Prove that if $(x,y) \in F$ and $x \leqslant 100$, then $y = 91$. (*Hint:* Translate into relational form and use induction on the stages F_k, with the induction hypothesis, "If $(x,y) \in F_j$ for some $j < k$, then....")

2.5 USING GLOBAL INFORMATION

The next thing we shall add to our model is the ability to use variables. It is easy enough to define the set of expressions with variables: If VAR is the set of variables, we just change line (i) of our definition of the set A in the previous section to read

(i) $\omega \cup \text{VAR} \subseteq A$

With the new definition, a variable may appear anywhere that an integer may. But now we can no longer define a function M: $A \rightarrow \omega$ assigning to each expression its value. Why? Let $\underline{\text{ALPHA}}$ and $\underline{\text{BETA}}$ be two members of VAR. Then at various times $M(\underline{(\text{ALPHA}) + (\text{BETA})})$ may have different values, depending on the values of $\underline{\text{ALPHA}}$ and $\underline{\text{BETA}}$.

So we need some additional information, which we shall call the *environment* in which $\underline{\text{ALPHA} + \text{BETA}}$ is being evaluated. What kind of information is needed in the environment? We need to know the value currently assigned to each variable. So we can think of an environment as a black box that, given as input the name of the variable, gives as output its current value (see Figure 2.5.1).

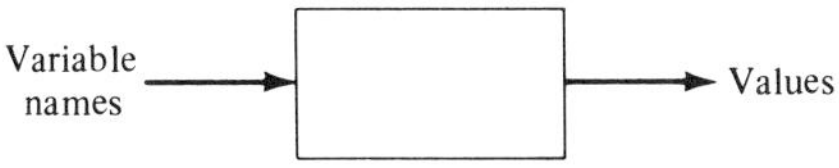

FIGURE 2.5.1 An environment.

Such a black box is just a function from variable names to values. In our example an environment is therefore a function I: VAR$\to\omega$. The set ENV of all environments is the set of all functions VAR$\to\omega$. We can now write down a definition of "addition expressions" A and a meaning function M: ENV$\times A\to\omega$, which gives the values of the expression in the given environment:

Definition of A.

(i) $\omega\subseteq A$
(ii) $\mathrm{VAR}\subseteq A$
(iii) if $x,y\in A$, then $\underline{(}x\underline{)}\,\underline{+}\,\underline{(}y\underline{)}\in A$
(iv) nothing else

Definition of M: ENV$\times A\to\omega$.

(i) $M(I,k)=k$ for $k\in\omega$
(ii) $M(I,v)=I(v)$ for $v\in\mathrm{VAR}$[9]
(iii) $M(I,\underline{(}x\underline{)}\,\underline{+}\,\underline{(}y\underline{)})=M(I,x)+M(I,y)$

Thus if $\underline{\text{ALPHA}}$, $\underline{\text{BETA}}\in\mathrm{VAR}$ and $I(\underline{\text{ALPHA}})=3$ and $I(\underline{\text{BETA}})=5$, then

$$
\begin{aligned}
M(I,\ \underline{\text{(ALPHA)+(BETA)}}\,) &= M(I,\ \underline{\text{ALPHA}}\,)+M(I,\ \underline{\text{BETA}}\,)\text{[by rule (iii)]}\\
&= I(\,\underline{\text{ALPHA}}\,)+\mathrm{I}(\,\underline{\text{BETA}}\,)\quad\text{[by rule (ii)]}\\
&=3+5\\
&=8
\end{aligned}
$$

We can do larger examples similarly. Here is a definition for expressions with addition, subtraction, and multiplication, using the usual procedure for infix arithmetic:

Let $V=\omega\cup\mathrm{VAR}\cup\{\underline{(}\,,\underline{)},\underline{+},\underline{*},\underline{-}\}$.

[9]Now we can see whether you understood footnote 8. In this line v is not underlined because it is a metavariable ranging over the variables, i.e., the members of the set VAR. We know this because the definition says "for $v\in\mathrm{VAR}$." Thus typical values of v are $\underline{\text{ALPHA}}$ and $\underline{\text{BETA}}$, as in the calculation following the definition.

	$F,T,E \subseteq V^+$	M_F, M_T, M_E: ENV $\times \begin{Bmatrix} F \\ T \\ E \end{Bmatrix} \rightarrow Z$
(i)	$\omega \subseteq F$	$M_F(I,k)=k,\ k \in \omega$
(ii)	VAR $\subseteq F$	$M_F(I,v)=I(v),\ v \in \mathrm{VAR}$
(iii)	if $t \in E$, then $\underline{(}\ t\ \underline{)} \in F$	$M_F(I,\ \underline{(}\ t\ \underline{)}\ )=M_E(I,t)$
(iv)	$F \subseteq T$	$M_T(I,t)=M_F(I,t),\ t \in F$
(v)	if $x \in T$ and $y \in F$, then $x \underline{*} y \in T$	$M_T(I,x \underline{*} y)=M_T(I,x) \times M_F(I,y)$
(vi)	$T \subseteq E$	$M_E(I,t)=M_T(I,t),\ t \in T$
(vii)	if $x \in E$ and $y \in T$, then $x \underline{+} y \in E$	$M_E(I,x \underline{+} y)=M_E(I,x)+M_T(I,y)$
(viii)	if $x \in E$ and $y \in T$, then $x \underline{-} y \in E$	$M_E(I,x \underline{-} y)=M_E(I,x)-M_T(I,y)$

So let VAR $=\{\underline{\mathrm{x}},\underline{\mathrm{y}}\}$, $I(\underline{\mathrm{x}})=3$, $I(\underline{\mathrm{y}})=4$:

$$\begin{aligned} M_E(I,\ \underline{\mathrm{x}}\ \underline{*}\ \underline{\mathrm{y}}) &= M_T(I,\ \underline{\mathrm{x}}\ \underline{*}\ \underline{\mathrm{y}}) \\ &= M_T(I,\ \underline{\mathrm{x}}) \times M_F(I,\ \underline{\mathrm{y}}) = M_F(I,\ \underline{\mathrm{x}}) \times I(\underline{\mathrm{y}}) \\ &= I(\underline{\mathrm{x}}) \times 4 = 3 \times 4 = 12 \end{aligned}$$

EXERCISES 2.5

1. Let VAR $=\{\underline{\mathrm{X}},\underline{\mathrm{Y}}\}$. Let B be defined by

   (i) VAR $\cup\, \omega \subseteq B$
   (ii) if $w, w' \in B$, then $\underline{(}w \underline{+} w'\underline{)} \in B$
   (iii) nothing else

   Let $M(I,t)$ be defined by

   (i) $M(I,v)=I(v)$ if $v \in \mathrm{VAR}$
   (ii) $M(I,k)=k$ if $k \in \omega$
   (iii) $M(I,\underline{(}w \underline{+} w'\underline{)})=M(I,w)+M(I,w')$
   (iv) nothing else

   Let $I=\{(\underline{\mathrm{X}},2),(\underline{\mathrm{Y}},3)\}$.
   Compute $M(I,\underline{((}5 \underline{+} \underline{X}\underline{)} \underline{+} \underline{(}\underline{\mathrm{Y}} \underline{+} 3\underline{)))}$.

2. Let VAR $=\{\underline{\mathrm{x}},\underline{\mathrm{y}}\}$ and let I be $\{(\underline{\mathrm{x}},3),(\underline{\mathrm{y}},5)\}$. Evaluate, using the definition of this section.

   (a) $M_E(I,4 \underline{*} \underline{\mathrm{x}} \underline{*} \underline{\mathrm{y}} \underline{+} \underline{\mathrm{x}} \underline{*} \underline{(}5 \underline{+} 3\underline{)})$
   (b) $M_E(I,5 \underline{-} 4 \underline{-} \underline{\mathrm{x}} \underline{*} 3)$
   (c) $M_E(I,5 \underline{-} \underline{(}4 \underline{-} \underline{\mathrm{x}} \underline{*} 3\underline{)})$

3 A Language for Programs

3.1 MATHEMATICAL DATA TYPES

We are now almost ready to embark on our project of using inductive techniques to define

$$\{w \in V^+ | w \text{ is a legal program in } L\}$$

and

$$\{(w,I,z) | w \text{ is a legal program in } L \text{ and } z \text{ is its output in environment } I\}$$

for some programming language L. We shall actually define a sequence of programming languages L, starting with a very simple language and ending with a very powerful one, but all of our languages will have in common the one thing that *all* programming languages have in common: they manipulate *data*. Hence we should stop and consider the nature of data.

Even in a simple language like FORTRAN, about the first thing one learns is that there are two data types: REAL and INTEGER. (Our languages will be even simpler: they will have only one data type.[1]) Clearly a data type is a set. But there is more to INTEGERs in FORTRAN than the set Z of integers:

(i) you can multiply, divide, add and subtract INTEGERs
(ii) you can compare two INTEGERs
(iii) you can write INTEGER constants

[1]This is an oversimplification, of course, since in practice one often needs more than one kind of data in the same program. We could solve this problem with additional notation, but we would not learn anything more as a result.

It is the ability to do these things that make INTEGERs useful in FORTRAN. By contrast, you can do very few things with the data type HOLLERITH, so HOLLERITH data are very difficult to manipulate in FORTRAN.

So, in fact, a data type is a set that you can do things with. To use more traditional mathematical language, it is a set that comes equipped with some "additional structure."[2] In general, we have a set A and

(i) some functions f: $A^n \rightarrow A$ for various n
(ii) some predicates p: $A^n \rightarrow \{\text{TRUE}, \text{FALSE}\}$ for various n
(iii) some constants $c \in A$

FORTRAN has only two-place INTEGER functions (and a one-place function if you count unary minus), but there is no reason for us to so restrict ourselves: We allow functions of n arguments for any n, and we might have (say) both 2- and 3-argument functions available. The same thing goes for predicates: FORTRAN INTEGERs use only two-place predicates (.LT.,.LE.,etc.) but our data types may have predicates of n arguments for any n. In FORTRAN, you can write down a constant to represent any INTEGER value, but again, we refuse to commit ourselves to this property: there need not be a constant for every member of the set A.[3]

We will write a data type as follows:

$$\mathcal{A} = \langle A, f_1, \ldots, f_n, p_1, \ldots, p_m, c_1, \ldots, c_r \rangle$$

where A is a nonempty set (called the *universe* or *carrier*), the f's are the functions, the p's are the predicates, and the c's are the constants.[4]

EXAMPLE 1

$$\mathcal{Z} = \langle Z, +, -, *, <, =, 0, 1 \rangle$$

Here are the integers, with conventional addition, subtraction, and multiplication, the predicates "less than" and "equal," and the constants 0 and

[2]The thought that you specify a structure by what you can do with it is the basic idea of something called *category theory*. Because of this connection, some of the purest of pure mathematics becomes relevant to computer science. Luckily, we need not get into this in order to reap its benefits.

What we are about to define is called in logic a *first-order structure*, [Shoenfield 67, Section 2.5].

[3]If, for example, you believe that FORTRAN REALs are just the real numbers, then there is no REAL constant equal to the value of the REAL expression (1./3.). Of course, the REALs are not the reals, any more than the INTEGERs are the integers. For a lucid discussion of the pitfalls of REAL arithmetic, see [Wirth 73, Section 8.4]. On the other hand, when *we* talk about the integers, we mean the integers: we are not going to worry about overflow. The analysis of roundoff error and how to keep it from ruining your day is the subject of the field called *numerical analysis* (see [Wilkes 66] or [Ralston 65]).

[4]Here is another bit of mathematical English that you may find confusing: Many books will say "where the f_i are the functions." Here "f_i" is a plural noun, meaning "the members of $\{f_i | i \in I\}$" (it could not mean $\{f_i | i \in I\}$, since this, being one set, is a singular noun). All in all, "the f's" seems to be the least confusing phrase.

1. Note that while there are only these two constants in $\mathfrak{Z}$, every $x \in Z$ is expressible as some expression involving 0, 1 and the operations of $\mathfrak{Z}$. This property will hold in most of our structures, though it need not hold in general.[5]

EXAMPLE 2

$$\mathfrak{R} = \langle \mathbb{R}, +, -, *, \div, <, =, 0, 1 \rangle$$

Here are the reals. Note that there are "unexpressible" values. One example is $\pi = 3.14159\ldots$.[6]

EXAMPLE 3

$$\mathfrak{N} = \langle \omega, +, \dot{-}, *, =0?, >, =, 0, 1, 2, \ldots \rangle$$

Here are the nonnegative integers with conventional addition and multiplication, and a constant for every element of ω. The two-place operation "$\dot{-}$," sometimes called "monus," is defined by

$$x \dot{-} y = \begin{cases} x - y & \text{if } x \geqslant y \\ 0 & \text{otherwise} \end{cases}$$

and $=0?$ is a one-place predicate given by $=0?(x) = \text{TRUE}$ iff $x = 0$.[7]

Here are some mathematical data types you may not have seen before:

EXAMPLE 4

$$\mathfrak{A} = \langle \{0, 1\}, +^{\mathfrak{A}}, *^{\mathfrak{A}}, 0^{\mathfrak{A}}, 1^{\mathfrak{A}} \rangle$$

Here we have put the superscript $\mathfrak{A}$ on our operations to indicate that these are like the old ones, but adapted to this particular structure. (We did something like this in creating "$\dot{-}$"). Here we mean $+^{\mathfrak{A}}$ and $*^{\mathfrak{A}}$ to be the two-place operations given by the following tables:

$+^{\mathfrak{A}}$	0	1
0	0	1
1	1	0

$*^{\mathfrak{A}}$	0	1
0	0	0
1	0	1

In mathematics, this is often called $\mathfrak{Z}_2$ or "the ring of integers modulo 2."[8]

[5]It is grossly false, in particular, for structures considered in mathematical logic, where such unexpressible elements play a crucial role.

[6]This is a deep theorem of algebra. It should not seem obvious. It is much easier to prove the existence of such unexpressible values than to prove that a particular real is unexpressible. Of course, one must be very careful about what one means by "expressible" (see [Herstein 64, Section 5.2]).

[7]*Quick*: Why can't we use plain minus? Notice that the name of the single predicate is the symbol "$=0?$." We shall often use question marks in the names of predicates.

[8]Because you can think of these operations as doing conventional addition or multiplication, then dividing by two and saving the remainder.

EXAMPLE 5. Let A be a set and let R be a relation on A, that is, $R \subseteq A \times A$. Define a predicate p_r: $A^2 \to \{\text{TRUE}, \text{FALSE}\}$ by

$$p_r(x,y) = \begin{cases} \text{TRUE} & \text{if } (x,y) \in R \\ \text{FALSE} & \text{otherwise} \end{cases}$$

Then we may talk about $\mathcal{Q} = \langle A, p_r \rangle$.

There are many data types that have arisen in computer science.

EXAMPLE 6

$$\mathcal{V} = \langle V^+, \circ, \preccurlyeq, v_1, \ldots, v_n \rangle$$

This data type[9] of *strings over* V has as its universe the set of strings over V. It has the single operation of concatenation. It has a single two-place predicate defined by

$$w \preccurlyeq v \qquad \text{iff} \quad w = v \text{ or there is some } x \in V^+ \text{ such that } wx = v$$

(this is read: w is an *initial segment* of v). The constants are the strings of one letter.

EXAMPLE 7. An example that we will use extensively is the data type of *binary stacks* (or just *stacks*). A stack is a list of zeros and ones. The operations are given as follows:

$$\mathbb{S} = \langle S, \text{push0}^{\mathbb{S}}, \text{push1}^{\mathbb{S}}, \text{pop}^{\mathbb{S}}, \text{is0?}^{\mathbb{S}}, \text{is1?}^{\mathbb{S}}, \text{empty?}^{\mathbb{S}}, \text{empty}^{\mathbb{S}} \rangle$$

$\text{push0}^{\mathbb{S}}$ and $\text{push1}^{\mathbb{S}}$ are one-place functions that push a 0 and a 1, respectively, onto the top of the stack which is its argument.

$\text{pop}^{\mathbb{S}}$ is a one-place function that removes the top element of its argument.

$\text{is0?}^{\mathbb{S}}$ and $\text{is1?}^{\mathbb{S}}$ are one-place predicates that are true iff the top element of the stack is a 0 or a 1, respectively.

$\text{empty?}^{\mathbb{S}}$ is a one-place predicate that is true iff the argument is the empty stack.

$\text{empty}^{\mathbb{S}}$ is the constant empty stack.

One may think of a stack as a stack of cafeteria trays, each of which is imprinted with either a 0 or a 1. One can operate only on the topmost tray in the stack. Figure 3.1.1 shows some of the functions on stacks.

$\text{pop}^{\mathbb{S}}(\text{empty}^{\mathbb{S}})$ causes some problems, since there is no element to be removed. We use the convention that $\text{pop}^{\mathbb{S}}(\text{empty}^{\mathbb{S}}) = \text{empty}^{\mathbb{S}}$.

For obvious typographical reasons, it is cumbersome to draw pictures like those of Figure 3.1.1 to represent stacks. Instead, we shall use strings

[9] Of course, this is only one of many possible choices for operations and predicates on V^+. The programming language SNOBOL4 (see [Griswold, Poage, and Polansky 71], [Pratt 75, Chapter 15]) contains many additional complex operations on strings.

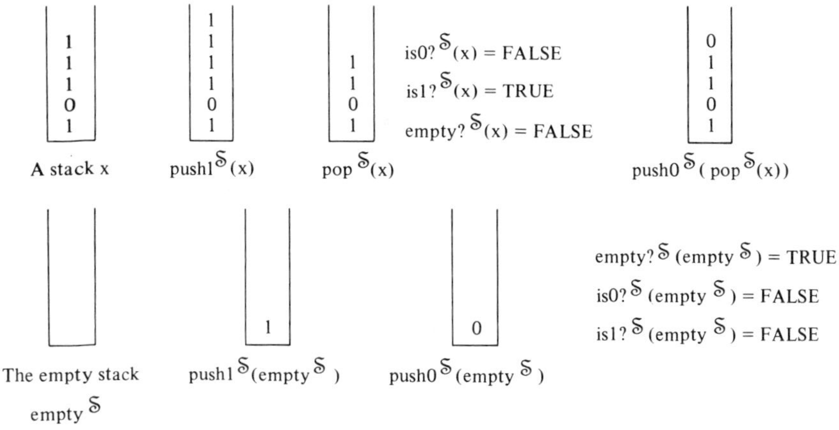

FIGURE 3.1.1

of 0's and 1's to represent stacks, with the left-hand end of the string representing the top (active) end of the stack. Thus for Figure 3.1.1 we have

$$\begin{aligned} x &= 11101 \\ \text{push1}^{\mathbb{S}}(x) &= 111101 \\ \text{pop}^{\mathbb{S}}(x) &= 1101 \\ \text{push0}^{\mathbb{S}}(\text{pop}^{\mathbb{S}}(x)) &= 01101 \end{aligned}$$

The empty stack is a problem in this representation. We shall write the empty stack (unsurprisingly) as "empty$^{\mathbb{S}}$."

The example of stacks will be used over and over to illustrate our language L, so it would be well to understand them and be able to calculate with them.

EXERCISES 3.1

In $\mathbb{S}$, find:

1. $\text{push1}^{\mathbb{S}}(\text{push0}^{\mathbb{S}}(\text{pop}^{\mathbb{S}}(\text{push1}^{\mathbb{S}}(\text{empty}^{\mathbb{S}}))))$

2. $\text{pop}^{\mathbb{S}}(\text{push1}^{\mathbb{S}}(\text{pop}^{\mathbb{S}}(\text{push0}^{\mathbb{S}}(\text{push0}^{\mathbb{S}}(\text{push1}^{\mathbb{S}}(\text{pop}^{\mathbb{S}}(\text{empty}^{\mathbb{S}})))))))$

In $\mathfrak{Z}$ as defined in the text, find:

3. $*(+(0,1), +(1, +(1,1)))$. (*Hint:* We are using prefix notation.)

4. $<(+(0,1), +(1, -(1,1)))$. (*Hint:* What kind of a function is $<$?)

In $\mathcal{Q}$ as defined in the text, find:

5. $*^{\mathcal{Q}}(+^{\mathcal{Q}}(0^{\mathcal{Q}}, 1^{\mathcal{Q}}), +^{\mathcal{Q}}(1^{\mathcal{Q}}, +^{\mathcal{Q}}(1^{\mathcal{Q}}, 1^{\mathcal{Q}})))$

3.2 LANGUAGES ON A DATA TYPE

Our pattern is to start with a data type $\mathcal{A}$ and produce a language L for manipulating $\mathcal{A}$. A language is of course a set of strings over some set of letters V. In this section we will construct V.

So let $\mathcal{A} = \langle A, f_1^{\mathcal{A}}, \ldots, f_n^{\mathcal{A}}, p_1^{\mathcal{A}}, \ldots, p_m^{\mathcal{A}}, c_1^{\mathcal{A}}, \ldots, c_r^{\mathcal{A}} \rangle$ be a data type. Our V will be the set consisting of the following members:

individual variable symbols: $\underline{\mathrm{x}}, \underline{\mathrm{y}}, \underline{\mathrm{z}}, \ldots, \underline{\mathrm{x}}_1, \underline{\mathrm{x}}_2, \underline{\mathrm{x}}_3, \ldots$;

function symbols: for each function $f_i^{\mathcal{A}}$, a function symbol $\underline{\mathrm{f}}_i$;

predicate symbols: for each predicate $p_j^{\mathcal{A}}$, a predicate symbol $\underline{\mathrm{p}}_j$;

constant symbols: for each constant $c_k^{\mathcal{A}}$, a constant symbol $\underline{\mathrm{c}}_k$;

punctuation symbols: $\underline{(}$, $\underline{)}$, and $\underline{,}$;

special symbols: depending on which language we are defining, we may add some special symbols, such as <u>if</u>, <u>then</u>, <u>else</u>, <u>begin</u>, and <u>end</u>.

We call $\underline{\mathrm{x}}, \underline{\mathrm{y}}, \underline{\mathrm{z}}, \ldots$ individual variable symbols because they will be used exclusively to denote individual elements of A.[10] The set of individual variable symbols is denoted IVS.

The members of the vocabulary V are often called the *elements* of the language.

It is worthwhile to recall at this point the distinction between objects and formal symbols. Each $\underline{\mathrm{f}}_i$ is a formal symbol—a *name*, like "Jane," "Frank," or "ALPHA." A name need not be attached to an object, or one object might have more than one name. We say that the name *denotes* the object to which it is attached. Of course, it will turn out that the name $\underline{\mathrm{f}}_i$ denotes the function $f_i^{\mathcal{A}}$ most of the time, but this is something we must make explicit.

For example, consider the data type of stacks

$$\mathbb{S} = \langle S, \mathrm{push0}^{\mathbb{S}}, \mathrm{push1}^{\mathbb{S}}, \mathrm{pop}^{\mathbb{S}}, \mathrm{is0?}^{\mathbb{S}}, \mathrm{is1?}^{\mathbb{S}}, \mathrm{empty?}^{\mathbb{S}}, \mathrm{empty}^{\mathbb{S}} \rangle$$

defined in Section 1. The set V for stacks consists of

$$\mathrm{IVS} \cup \{ \underline{\mathrm{push0}}, \underline{\mathrm{push1}}, \underline{\mathrm{pop}}, \underline{\mathrm{is0?}}, \underline{\mathrm{is1?}}, \underline{\mathrm{empty?}}, \underline{\mathrm{empty}} \}$$
$$\cup \{ \underline{(}, \underline{)}, \underline{,} \} \cup \{\text{special symbols as needed}\}$$

3.3 THE LANGUAGE OF TERMS

At long last we are ready to define our first language L. We shall use our new convention, so we are working in the data type

$$\mathcal{A} = \langle A, f_1^{\mathcal{A}}, \ldots, f_r^{\mathcal{A}}, p_1^{\mathcal{A}}, \ldots, p_m^{\mathcal{A}}, c_1^{\mathcal{A}}, \ldots, c_r^{\mathcal{A}} \rangle$$

[10]Rather than, say, subsets of A or functions $A \to A$.

We shall define a language $L_{\mathcal{Q}}$ (we will write just L), and a function M: ENV$\times L \to A$ that takes an environment (again, an environment is a function I: IVS$\to A$ that gives the current value of each individual variable) and a string in the language and gives as its value the "value" of the string in that environment.

Our first language is the language of *terms*. Its construction is similar to the construction of FORTRAN arithmetic expressions in Section 8. It is simpler because we do not worry about precedence.

	Terms	"Meaning" function M
(i)	if c is a constant symbol, then c is a term	$M(I,c)=c^{\mathcal{Q}}$
(ii)	if v is an individual variable symbol, then v is a term	$M(I,v)=I(v)$
(iii)	if f is an n-place function symbol and $t_1,\ldots,t_n$ are terms, then $f\underline{(}t_1\underline{,}\ \cdots\ \underline{,}\ t_n\underline{)}$ is a term	$M(I,f\underline{(}t_1\underline{,}\ \cdots\ \underline{,}\ t_n\underline{)})= f^{\mathcal{Q}}(M(I,t_1),\ldots,M(I,t_n))$
(iv)	nothing else	

Here we have again underscored the punctuation in the definition of a term to emphasize that they are formal symbols, like punches in a card.[11]

The language of terms we have just defined is just the conventional mathematical set of terms: If $\underline{\mathrm{f}}$, $\underline{\mathrm{g}}$, and $\underline{\mathrm{h}}$ are two-place function symbols, and $\underline{\mathrm{a}}$ is a constant symbol, then a typical term is $\underline{\mathrm{f(g(x,y),h(a,g(y,z)))}}$.

If we have some term t, what is $M(I,t)$? If t is a constant symbol $\underline{\mathrm{c}}_k$, then $M(I,t)$ is just $c_k^{\mathcal{Q}}$. (Thus the constant symbol $\underline{\mathrm{c}}_k$ denotes the object $c_k^{\mathcal{Q}} \in A$.) If t is just an individual variable symbol $\underline{\mathrm{x}}_i$, then the value $M(I,t)$ of t is just $I(\underline{\mathrm{x}}_i)$, the current value of $\underline{\mathrm{x}}_i$. If t is neither of these, it must be of the form $f_i\underline{(}t_1\underline{,}\ldots\underline{,}\, t_n\underline{)}$ where the t_i are terms. In that case $M(I,t)$ is computed as follows: Calculate $M(I,t_1),\ldots,M(I,t_n)$ and apply $f_i^{\mathcal{Q}}$ (the function $A^n \to A$, not the formal symbol $\underline{\mathrm{f}}_i$) to the results.[12]

[11]Now let us review our discussion of language versus metalanguage, which we carried on in Chapter 1, footnote 11, and Chapter 2, footnotes 8 and 9, and see whether we can account for the underlining or lack of it on the various symbols in this definition. First of all, c, v, and f are metavariables, ranging over constant symbols, individual variable symbols, and function symbols. Similarly, $t_1,\ldots,t_n$ are metavariables ranging over terms. In clause (iii) of the definition of a term, the parentheses and commas are underscored because they are the actual symbols that will appear in the string. In clause (iii) of the definition of M, the parentheses and commas are underscored on the left-hand side for the same reason; on the right-hand side they are not underscored because they are the conventional grouping symbols of the metalanguage. The periods (ellipses) are never underlined because they never appear in a term. Last, $f^{\mathcal{Q}}$ is the operation in $\mathcal{Q}$ corresponding to the function symbol f. So if f is $\underline{\mathrm{push0}}$, then $f^{\mathcal{Q}}$ is the function push0$^{\mathcal{S}}$.

[12]In other words: You just evaluate t as if it were an ordinary mathematical expression. But you can't evaluate a formal symbol such as $\underline{\mathrm{x}}$ or $\underline{\mathrm{f}}$ as if it were 3 or π. So what do you do?

So for the type $\mathbb{S}$ of stacks,

$$M(I, \underline{\text{push1(push0(x))}}\,) = \text{push1}^{\mathbb{S}}(M(I, \underline{\text{push0(x)}}\,))$$
$$= \text{push1}^{\mathbb{S}}(\text{push0}^{\mathbb{S}}(M(I, \underline{\text{x}}\,)))$$
$$= \text{push1}^{\mathbb{S}}(\text{push0}^{\mathbb{S}}(I(\underline{\text{x}}\,)))$$

This says: Whatever the current value of $\underline{\text{x}}$ is, push a 0 onto it, push 1 onto that, and the result is your answer. If $I(\underline{\text{x}}) = 110$, then

$$M(I, \underline{\text{push1(push0(x))}}\,) = \text{push1}^{\mathbb{S}}(\text{push0}^{\mathbb{S}}(I(\underline{\text{x}}\,))) = 10110$$

Notice that this has no effect on the current value of $\underline{\text{x}}$. We have not redefined the function I![13]

We shall be using this meaning function for terms throughout the book; since we always use M for meaning functions, we distinguish this one by calling it M_{terms}.

Another example: let $I(\underline{\text{x}}) = 110$,

$$M(I, \underline{\text{push0(pop(push1(push0(x))))}}\,)$$
$$= \text{push0}^{\mathbb{S}}(M(I, \underline{\text{pop(push1(push0(x)))}}\,))$$
$$= \text{push0}^{\mathbb{S}}(\text{pop}^{\mathbb{S}}(M(I, \underline{\text{push1(push0(x))}}\,)))$$
$$\vdots$$
$$= \text{push0}^{\mathbb{S}}(\text{pop}^{\mathbb{S}}(\text{push1}^{\mathbb{S}}(\text{push0}^{\mathbb{S}}(I(\underline{\text{x}}\,)))))$$
$$\vdots$$
$$= \text{push0}^{\mathbb{S}}(\text{pop}^{\mathbb{S}}(10110))$$
$$= \text{push0}^{\mathbb{S}}(0110)$$
$$= 00110$$

Notice how a calculation is just a chain of applications of the identities that define M. A term is thus very similar to a straight-line program: that is, a piece of program without branching. In our next language we introduce branching.

EXERCISES 3.3

In the data type $\mathbb{S}$, define the environments I and J by

$$I = \{(\underline{\text{x}}, \text{empty}^{\mathbb{S}}), (\underline{\text{y}}, 1)\}$$
$$J = \{(\underline{\text{x}}, 0), (\underline{\text{y}}, 11)\}$$

If you hit an individual variable symbol like $\underline{\text{x}}$, you use the environment I to evaluate it. If you hit a function symbol $\underline{\text{f}}_i$, pretend it is $f_i^{\mathcal{a}}$.

[13] So a more accurate description of $M(I, \underline{\text{push1(push0(x))}}\,)$ is this: Take the stack that is the current value of $\underline{\text{x}}$. *Imagine* the stack you *would* obtain if you pushed a 0 and then a 1 onto the stack. That stack is the answer.

Evaluate (using the definition) the following terms over the data type $\mathbb{S}$ in the specified environment. Be sure to show your work.

1. push1(push0(x)) in the environment I.

2. push1(push0(x)) in the environment J.

3. push0(push1(pop(push1(push0(pop(y)))))) in J.

4. pop(pop(pop(pop(push1(empty))))) in I.

Let $K=\{(\underline{x},2),(\underline{y},3),(\underline{z},5)\}$. Evaluate in $\mathfrak{Z}$:

5. $+(1,+(1,-(z,y)))$ in K.

6. Evaluate the following term in $\mathfrak{N}$ and $\mathfrak{C}$ (as defined in Section 2.1, Example 4) at $I(\underline{x})=0$, $I(\underline{y})=1$:

$$*(+(x,+(y,y)),+(1,*(y,y)))$$

3.4 THE LANGUAGE OF CONDITIONALS

The language of conditionals on a data type $\mathfrak{C}$ is a simple extension of the language of terms over $\mathfrak{C}$. We introduce three new special symbols: if, then, and else. Here is the definition of the set of conditionals and our new meaning function, which we again call M:

Conditionals

(i) every term is a conditional
(ii) if p is an n-place predicate symbol, and $u_1,\ldots,u_n,t_1,t_2$ are contitionals, then

$$\underline{\text{if}}\ p(u_1,\ldots,u_n)\ \underline{\text{then}}\ t_1\ \underline{\text{else}}\ t_2$$

is a conditional
(iii) nothing else

M

(i) if t is a term, $M(I,t)=M_{\text{terms}}(I,t)$
(ii-t) if $p^{\mathfrak{C}}(M(I,u_1),\ldots,M(I,u_n))=\text{TRUE}$, then

$$M(I,\underline{\text{if}}\ p(u_1,\ldots,u_n)\ \underline{\text{then}}\ t_1\ \underline{\text{else}}\ t_2)=M(I,t_1)$$

(ii-f) if $p^{\mathfrak{C}}(M(I,u_1),\ldots,M(I,u_n))=\text{FALSE}$, then

$$M(I,\underline{\text{if}}\ p(u_1,\ldots,u_n)\ \underline{\text{then}}\ t_1\ \underline{\text{else}}\ t_2)=M(I,t_2)$$

Some examples of conditionals:

push0(pop(x))

if is0?(x) then push1(x) else y

if is0?(pop(x)) then push1(x) else if is1?(x) then push0(x) else pop(y)

Let us analyze the last example to check that it fits under (ii):

the predicate symbol p is is0?

the conditional u_1 is the term pop(x), which is a conditional by (i)

the conditional t_1 is the term push1(x), which is a conditional by (i)

the conditional t_2 is if is1?(x) then push1(x) else pop(y), which we can analyze similarly.

Here u_1 and t_1 were just terms, but in general they could be other conditionals:

if is0?(if is1?(x) then y else pop(y))
then if is1?(y) then pop(y) else y
else pop(pop(y))

Here

u_1 is the conditional	if is1?(x) then y else pop(y)
t_1 is the conditional	if is1?(y) then pop(y) else y
t_2 is the conditional	pop(pop(y))

We sometimes refer to t_1 as the *then-part* and to t_2 as the *else-part*. It is sometimes helpful to draw pictures showing how a conditional is broken up into parts (see Figure 3.4.1).[14]

Again, we evaluate a conditional exactly the way one would expect. If it is just a term, we use the meaning function for terms, M_{terms}, defined in Section 3. Otherwise, to evaluate if $p(u_1,\ldots,u_n)$ then t_1 else t_2 in environment I, we evaluate $u_1,\ldots,u_n$, getting $M(I,u_1),\ldots,M(I,u_n)$ [since that is what $M(I,u_i)$ is], and check to see whether $p^{\mathcal{Q}}$ is true for this n-tuple of values. If it is true, the value of the conditional is the value of t_1. If it is false, the value of the conditional is the value of t_2.

Let us see what this implies for the chain-identity picture. If, while evaluating a term, we reach an occurrence of $M(I,t)$, we can tell which

[14]This business of taking a string, determining whether or not it is in an inductively defined set, and, if so, analyzing it to find its components (in this case u_1, t_1, and t_2), is called *parsing*. For a discussion of parsing algorithms and related topics see [Prather 76, Chapter 6] or [Denning, Dennis, and Qualitz 78].

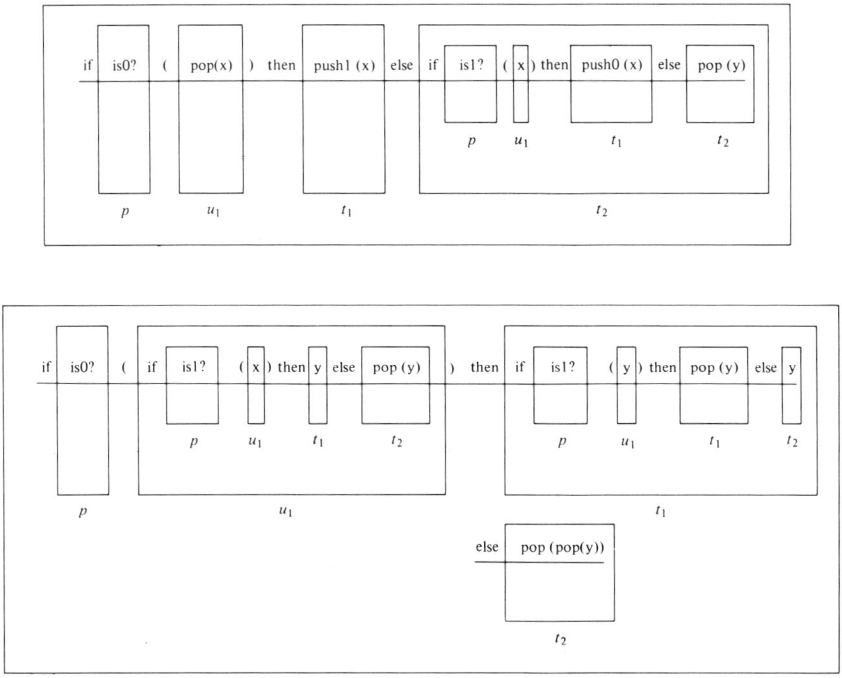

FIGURE 3.4.1 Diagrams for decomposing two conditionals.

identity to apply merely by looking at t. For a conditional, it may take more work to determine which identity is applicable. We first evaluate the if-part (which may alone require a long computation). If the answer is TRUE, then (ii-t) applies; if the answer is FALSE, then (ii-f) applies. We refer to the evaluation of the if-part as a *subsidiary calculation* and indent it when we present the main calculation.

EXAMPLE 1. Let $I(\underline{x}) = 10111$ (remember, the top of the stack is at the left-hand end). Let $t = \underline{\text{if is0?(pop(x)) then push1(x) else push0(x)}}$. Evaluate $M(I,t)$ as follows:

$$M(I,\ \underline{\text{if is0?(pop(x)) then push1(x) else push0(x)}}\ )$$

$\text{is0?}^{\$}(M(I,\ \underline{\text{pop(x)}}\ ))$
$= \text{is0?}^{\$}(\text{pop}^{\$}(M(I,\ \underline{\text{x}}\ )))$
$= \text{is0?}^{\$}(\text{pop}^{\$}(10111))$
$= \text{is0?}(0111)$
$= \text{TRUE}$

$$
\begin{aligned}
&= M(I,\ \underline{\text{push1(x)}}\ ) \\
&= \text{push1}^{\mathcal{S}}(M(I,\ \underline{\text{x}}\ )) \\
&= \text{push1}^{\mathcal{S}}(10111) \\
&= 110111
\end{aligned}
$$

If $I(\underline{\text{x}}) = 0100$, then

$$M(I,\ \underline{\text{if is0?(pop(x)) then push1(x) else push0(x)}}\ )$$

$$
\boxed{
\begin{aligned}
&\text{is0?}^{\mathcal{S}}(M(I,\ \underline{\text{pop(x)}}\ )) \\
&\qquad= \text{is0?}^{\mathcal{S}}(\text{pop}^{\mathcal{S}}(M(I,\ \underline{\text{x}}\ ))) \\
&\qquad= \text{is0?}^{\mathcal{S}}(\text{pop}^{\mathcal{S}}(0100)) \\
&\qquad= \text{is0?}^{\mathcal{S}}(100) \\
&\qquad= \text{FALSE}
\end{aligned}
}
$$

$$
\begin{aligned}
&= M(I,\ \underline{\text{push0(x)}}\ ) \\
&= \text{push0}^{\mathcal{S}}(M(I,\ \underline{\text{x}}\ )) \\
&= \text{push0}^{\mathcal{S}}(0100) \\
&= 00100
\end{aligned}
$$

Note that evaluating $\underline{\text{pop(x)}}$ did *not* change the value of $\underline{\text{x}}$ that was subsequently used to evaluate $\underline{\text{push1(x)}}$ or $\underline{\text{push0(x)}}$.

EXAMPLE 2. Let $I(\underline{\text{x}}) = 111$:

$$M(I,\ \underline{\text{if is0?(x) then x else if is0?(pop(x)) then pop(x) else pop(pop(x))}}\ )$$

$$
\boxed{
\begin{aligned}
&\text{is0?}^{\mathcal{S}}(M(I,\ \underline{\text{x}}\ )) \\
&\qquad= \text{is0?}^{\mathcal{S}}(111) \\
&\qquad= \text{FALSE}
\end{aligned}
}
$$

$$= M(I,\ \underline{\text{if is0?(pop(x)) then pop(x) else pop(pop(x))}}\ )$$

$$
\boxed{
\begin{aligned}
&\text{is0?}^{\mathcal{S}}(M(I,\ \underline{\text{pop(x)}}\ )) \\
&\qquad= \text{is0?}^{\mathcal{S}}(\text{pop}^{\mathcal{S}}(M(I,\ \underline{\text{x}}\ ))) \\
&\qquad= \text{is0?}^{\mathcal{S}}(\text{pop}^{\mathcal{S}}(111)) \\
&\qquad= \text{is0?}^{\mathcal{S}}(11) \\
&\qquad= \text{FALSE}
\end{aligned}
}
$$

$= M(I, \underline{\text{pop(pop(x))}})$
$= \text{pop}^{\mathbb{S}}(M(I, \underline{\text{pop(x)}}))$
$= \text{pop}^{\mathbb{S}}(\text{pop}^{\mathbb{S}}(M(I, \underline{\text{x}})))$
$= \text{pop}^{\mathbb{S}}(\text{pop}^{\mathbb{S}}(111))$
$= \text{pop}^{\mathbb{S}}(11)$
$= 1$

Again: There is nothing subtle about evaluating a conditional expression: the formal symbols $\underline{\text{if}}$, $\underline{\text{then}}$, and $\underline{\text{else}}$ are interpreted as the usual "if," "then," and "else"; the individual variable symbols are interpreted using I, and the function and predicate symbols are interpreted using the data type $\mathcal{A}$.

EXERCISES 3.4

1. Show, using the definition, that each of the strings in Exercises 4, 5, 6 are in fact conditionals.

2. Why is $\underline{\text{push0(if is0?(x) then x else y)}}$ not a conditional?

3. In Example 1, why did we write $\text{is0?}^{\mathbb{S}}(M(I, \underline{\text{pop(x)}}))$ instead of $M(I, \underline{\text{is0?(pop(x))}})$?
   Let

$$I = \{(\underline{\text{x}}, 101), (\underline{\text{y}}, 001)\}$$
$$J = \{(\underline{\text{x}}, 011), (\underline{\text{y}}, 10)\}$$
$$K = \{(\underline{\text{x}}, 00), (\underline{\text{y}}, 0)\}$$
$$L = \{(\underline{\text{x}}, 0), (\underline{\text{y}}, 110)\}$$

   In the data type $\mathbb{S}$ of stacks, evaluate each of the following conditional expressions in each of the four environments listed above.

4. $\underline{\text{if is1?(x) then push1(x) else if is1?(pop(x)) then push0(x) else push1(pop(y))}}$

5. $\underline{\text{if is1?(x) then if is1?(y) then push1(pop(x)) else push0(pop(x)) else if is1?(y)}}$ $\underline{\text{then push0(pop(y)) else y}}$

6. $\underline{\text{if is1?(pop(x)) then x else if is0?(pop(x)) then y else if is1?(y) then push0(pop(y))}}$ $\underline{\text{else pop(y)}}$

3.5 THE LANGUAGE OF EXPRESSIONS

We now have a language, called the language of conditionals, for manipulating the data type $\mathcal{A}$. It allows us straight-line composition and branching, but it does not have loops, subprograms, or other things that we conventionally associate with a programming language. In this section we shall add the feature of *function subprograms*. It turns out that of all the features we might wish to add, function subprograms are the easiest and the most powerful: Using subroutines[15] we can get the effects of loops, labels, and even assignment statements.[16] We call the strings in this language *expressions*.

If we are to have programmer-supplied functions in our language, we must at the very least have names for the functions to be defined. Hence we add to the elements of the language a set FVS of *function variable symbols*. These are symbols for programmer-written functions, not to be confused with the set IVS of individual variable symbols. Why do we need a different set of symbols? In the first place, we have to be able to tell the difference between a variable and a subroutine call,[17] and this is the easiest way. Furthermore, it is necessary to know how many arguments to supply to a subroutine, just as we must know how many arguments a built-in function $f_i^@$ takes. So we associate with FVS a map r: FVS$\to\omega$, called the "ranking map," which, given a symbol F in FVS, tells how many arguments F takes.[18] If $r(F)=n$, we say F is an *n-place function variable*

[15]We shall use the word "subroutine" to mean what is called in FORTRAN a *function program*, in ALGOL a (typed) *procedure*, or in PASCAL a *function*—that is, a subprogram that takes some arguments and returns a value. It is on the property of returning a value that we shall focus.

A *function subprogram* or ALGOL procedure may do things other than return a value. For example, consider

```
      FUNCTION ICNT (IA,N)
      DIMENSION IA(N)
      ICNT=0
      DO 10 I=1,N
      IF (IA(I) .NE. 0) ICNT=ICNT+1
      IA(I)=0
   10 CONTINUE
      RETURN
      END
```

If this function is called, say, by

```
J=ICNT(IB,100)
```

then the first 100 locations of array IB will be set to zero, and the number of nonzero locations found will be returned and assigned to J. Such "other things" as changing the values of main program variables are called *side effects*. Uncontrolled side effects have the primary effect of making programs harder to read (who, after all, would expect the modest little assignment statement above to wreak such havoc with array IB?). We get along quite well without them.

[16]See Section 7.

[17]After all, every programming language has to do this.

[18]This is all to say that you can tell how many arguments a subroutine takes just by looking at its *name*. This is not quite how real programming languages work: A given function name might be used for functions of different numbers of arguments in different programs. We could fix this, but the difference is so slight it hardly matters. The business of building an exact model of a particular language, with all of these minor discrepancies removed, is called *programming language semantics*. Such exact models are important in standardizing programming languages.

symbol. The ranking map is part of the declaration of a set of function variable symbols, just as the domain and range were part of the declaration of a function.

Again, the introduction of subroutines requires the introduction of a new class of global or environmental information. The value of an expression depends not only on the current values of the individual variables, but also on the subroutines associated with the function variable symbols appearing in the expression. So this portion of the environment may be represented by a map[19] δ: FVS→Exp which sends each function variable symbol (fvs for short) to the expression. We call this the *functional environment*.

That is, if F is an n-place fvs, then writing $\delta(F)=t$ is the same as writing

in LISP: (DEFINE $F(X_1,\ldots,X_n)\ t$)
or in FORTRAN: FUNCTION $F(X_1,\ldots,X_n)$
$F=t$
RETURN
END

Here F is the function *name*, t is the function *body*, and $X_1,\ldots,X_n$ are the *formal parameters*.

Now, in order to evaluate an expression t, we need three classes of "environmental information":

1. the structure $\mathcal{A}$: it tells how to interpret function and predicate symbols and constants ("built-in functions")
2. the environment I: it tells how to interpret individual variable symbols
3. the functional environment δ: it tells how to interpret the function variable symbols ("programmer-defined functions")

Since we are writing a language for a single data type $\mathcal{A}$, we will not list $\mathcal{A}$ as an argument of our meaning function M. So a typical call on the meaning function for expressions will look like[20]

$$M(\delta,I,t)$$

The astute reader will note that we have not yet defined the set of expressions. This is in contrast with our practice in the last two sections, where we started with the definition and then proceeded with the explanation. Here the procedure is reversed, for several good reasons. First, we have introduced two new things to the language: the set FVS and the new functional environment δ. Second, this procedure more accurately reflects the way formal systems are constructed. We first decide what we want the system to do—in this case, subroutines. We then consider what kind of

[19]This is the lower case Greek letter *delta*.

[20]The suppression of $\mathcal{A}$ is in keeping with our previous practices. Strictly speaking, we are defining a new M for each $\mathcal{A}$; if we put in $\mathcal{A}$ as a parameter to M we would have a hard time figuring out what the domain of M was. Later on we shall also suppress δ.

information it will be necessary to keep—in this case, FVS and δ.[21] We then see what kinds of things can cause trouble, and try to "debug" the system to avoid them.

There is, in fact, at least one bug in the scheme we are talking about. Normally, the body of a subroutine only involves the variables that are formal parameters.[22] In particular, a subroutine may not refer to variables in the main program. In our system, if F is an n-place fvs, then δF, the body of F,[23] may contain only n variables. So not every δ is a reasonable map for an environment.

We therefore make the following restriction:

Definition. The set FENV of functional environments is defined to be $\{\delta \mid \delta\colon \text{FVS} \to \text{Exp}$ and if $r(F)=n$, then δF contains no individual variable symbols other than $\underline{x}_1,\ldots,\underline{x}_n\}$.

Note that this definition will not really make sense until we have defined Exp.

We are now, at long last, able to write down the definition of the set Exp of expressions and the evaluation function M in Table 3.5.1.

Let us see what all this means. On the left-hand side we have an inductive definition of a set of strings called *expressions*. An expression is anything built from individual variables (i) and constant symbols (ii) by

constructing terms (iii),
constructing conditionals (iv), and
constructing subroutine calls (v).

Every expression is in exactly one of the five classes,[24] so to evaluate an expression we need only specify how to evaluate each class.

Once the set Exp of expressions is defined, we can define the set FENV of functional environments using the preceding definition. We may then define the function M: FENV$\times$ENV$\times$Exp$\to A$ (recall A is the set of values of the data type $\mathcal{Q}$).

Now we can consider how the evaluation function M works. The first four cases are just as before. Individual variables and constants are evaluated using $\mathcal{Q}$ and I, as usual. To evaluate a term $f_i(t_1,\ldots,t_n)$, you evaluate $t_1,\ldots,t_n$ and apply $f_i^{\mathcal{Q}}$ to the results. To evaluate a conditional,

[21] The astute reader will also notice the parallel to program writing.

[22] Since we do not have assignments, it will turn out that local variables are unnecessary, and COMMON blocks need special (and complicated) treatment so we will ignore them. Similarly, we do not have any block structure (as in ALGOL, PASCAL, or PL/I) so nonlocal variables make no sense.

[23] We shall often write δF for $\delta(F)$.

[24] And is in fact unambiguous, so M is a partial function.

TABLE 3.5.1 The language of expressions over the data type $\mathcal{A}$

	Definition of Exp		Definition of M: FENV $\times$ ENV $\times$ Exp $\to A$
(1)	IVS $\subseteq$ Exp	(1)	$M(\delta, I, v) = I(v)$
(2)	If c is a constant symbol, then $c \in$ Exp	(2)	$M(\delta, I, c) = c^{\mathcal{A}}$
(3)	If f is an n-place function symbol, and $t_1, \ldots, t_n \in$ Exp, then $\underline{f}\underline{(}t_1\underline{,} \ldots \underline{,} t_n\underline{)} \in$ Exp	(3)	$M(\delta, I, \underline{f}\underline{(}t_1\underline{,} \ldots \underline{,} t_n\underline{)})$ $= f^{\mathcal{A}}(M(\delta, I, t_1), \ldots, M(\delta, I, t_n))$
(4)	If p is an n-place predicate symbol, and $u_1, \ldots, u_n, t_1, t_2 \in$ Exp, then $\underline{\text{if}}\ \underline{p}\underline{(}u_1\underline{,} \ldots \underline{,} u_n\underline{)}\ \underline{\text{then}}\ t_1\ \underline{\text{else}}\ t_2 \in$ Exp	(4-t)	If $p^{\mathcal{A}}(M(\delta, I, u_1), \ldots, M(\delta, I, u_n)) =$ TRUE then $M(\delta, I,\ \underline{\text{if}}\ \underline{p}\underline{(}u_1\underline{,} \ldots \underline{,} u_n\underline{)}$ $\underline{\text{then}}\ t_1\ \underline{\text{else}}\ t_2) = M(\delta, I, t_1)$
		(4-f)	If $p^{\mathcal{A}}(M(\delta, I, u_1), \ldots, M(\delta, I, u_n)) =$ FALSE then $M(\delta, I,\ \underline{\text{if}}\ \underline{p}\underline{(}u_1\underline{,} \ldots \underline{,} u_n\underline{)}$ $\underline{\text{then}}\ t_1\ \underline{\text{else}}\ t_2) = M(\delta, I, t_2)$
(5)	If F is an n-place fvs, and $u_1, \ldots, u_n \in$ Exp, then $\underline{F}\underline{(}u_1\underline{,} \ldots \underline{,} u_n\underline{)} \in$ Exp	(5a)	"Call-by-Value" Let $I'(\underline{x}_i) = M(I, u_i)$ for $i = 1, \ldots, n$ then $M(\delta, I, \underline{F}\underline{(}u_1\underline{,} \ldots \underline{,} u_n\underline{)}) = M(\delta, I', \delta F)$
		(5b)	"Call-by-Name" $M(\delta, I, \underline{F}\underline{(}u_1\underline{,} \ldots \underline{,} u_n\underline{)}) = M\left(\delta, I, \delta F\begin{bmatrix} u_1 & \cdots & u_n \\ \underline{x}_1 & \cdots & \underline{x}_n \end{bmatrix}\right)$
(6)	Nothing else	(6)	Nothing else

you evaluate the test, and then the value is either the value of the "then-clause" t_1 or the value of the "else-clause" t_2.

What is really new here is the feature of subroutine calls. It turns out that there are two quite distinct ways of handling these calls, called *call-by-value* and *call-by-name*. We therefore have two distinct meaning functions M, one obtained by using Table 3.5.1 with rule (5a), and the second obtained by using Table 3.5.1 with rule (5b). In any example, we will always specify which of these two functions is to be used.

The first, called call-by-value, is inspired by the way FORTRAN passes parameters.[25] To evaluate the following code,

```
Z=F(1.0,2.0)
⋮
FUNCTION F(X,Y)
F=X+Y
END
```

the function body of F is started, with the value of X set to 1.0 and the value of Y set to 2.0. In other words, we create a new environment with $I(\mathrm{X})=1.$ and $I(\mathrm{Y})=2.$, and then evaluate the body of F in this new environment.

So in our system, to evaluate $M(I,F(t_1,\ldots,t_n))$ using call by value, we first create a new environment I' with $I'(\underline{\mathrm{x}}_1)=M(I,t_1),\ldots,I'(\underline{\mathrm{x}}_n)=M(I,t_n)$. Here $t_1,\ldots,t_n$ are called the *actual parameters*. Thus in the new environment I', the value of $\underline{x}_i$ will be the value of t_i. Then, in the new environment I', we evaluate δF, the body of F. The construction of I' is a subsidiary calculation in the chain-identity picture, just like the evaluation of the predicate in a conditional. Since δ never changes, we drop it in our calculations.

EXAMPLE 1

$$\delta\,\underline{\mathrm{F}} = \underline{\text{if is0?}(\mathrm{x}_1)\text{ then push0}(\mathrm{x}_1)\text{ else push1}(\mathrm{x}_1)}$$

$$I(\underline{\mathrm{x}_1})=011$$

$$t=\underline{\mathrm{F}(\text{push1}(\mathrm{x}_1))}$$

To evaluate $M(I,t)$ we proceed as follows:

$$M(I,\ \underline{\mathrm{F}(\text{push1}(\mathrm{x}_1))}\ ) \qquad (1)$$

$$I'(\underline{\mathrm{x}}_1)=M(I,\ \underline{\text{push1}(\mathrm{x}_1)})=\text{push1}^{\mathrm{S}}(M(I,\ \underline{\mathrm{x}})) \qquad (2)$$

$$=\text{push1}^{\mathrm{S}}(011)=1011 \qquad (3)$$

[25]Actually, to account for the side effects in FORTRAN, one needs a much more complex mechanism, called *call-by-reference*.

$$= M(I', \underline{\text{if is0?}(\text{x}_1) \text{ then push0}(\text{x}_1)\text{else push1}(\text{x}_1)}\ ) \quad (4)$$

$$\boxed{\text{is0?}^{\$}(M(I',\underline{\text{x}}_1)) = \text{is0?}^{\$}(1011) = \text{FALSE}} \quad (5)$$

$$= M(I', \underline{\text{push1}(\text{x}_1)}\ ) \quad (6)$$

$$= \text{push1}^{\$}(M(I', \underline{\text{x}}_1)) \quad (7)$$

$$= \text{push1}^{\$}(1011) \quad (8)$$

$$= 11011 \quad (9)$$

So 11011 is the answer. Note that in line (7) the value of $\underline{\text{x}}_1$ used was 1011, its value in the subroutine, and not 011, its value in the main program. Note also that if F is an n-ary function variable symbol, the only variables appearing in δF are $\underline{\text{x}}_1,\ldots,\underline{\text{x}}_n$, and so we need only specify I' for these variables. Luckily,[26] that is precisely what Table 3.5.1 gives.

A second method of handling subroutines is inspired by the "macro" feature found in many assembly languages. In these languages, if one writes

```
MACRO ADDUP (X,Y)
CLA X
ADD Y
END
  ⋮
ADDUP (A+2,B+3)
  ⋮
```

the code which is produced is

```
  ⋮
CLA A+2
ADD B+3
  ⋮
```

In other words, the body of the subroutine is produced, with a copy of the correct argument (the actual parameter) substituted for each occurrence of a variable (the formal parameter) in the body. This is referred to as *call-by-name*.

So in our system, to evaluate $M(I,F(\underline{t_1,\ldots,t_n}))$ using call-by-name, we substitute t_i for each occurrence of $\underline{\text{x}}_i$ in δF, and evaluate the resulting expression[27] in the original environment I. The variables that appear in the substituted expression are the "main program variables"—all of the subroutine's variables (the formal parameters) are wiped out by the substitu-

[26]Well, actually more by design than by luck.

[27]This process is called *thunking* (believe it or not). See [Pratt 75].

tion.[28] Therefore the original environment is the correct one. We use brackets to denote substitution, and write

$$M(I, F(t_1, \ldots, t_n)) = M\left(I, \delta F\begin{bmatrix} t_1 & \cdots & t_n \\ x_1 & \cdots & x_n \end{bmatrix}\right)$$

Let us do the last example with call-by-name:

EXAMPLE 2. Again

$$\delta F = \text{if is0?}(x_1) \text{ then push0}(x_1) \text{ else push1}(x_1)$$

$$I(x_1) = 011$$

$$t = F(\text{push1}(x_1))$$

$$M(I, F(\text{push1}(x_1)))$$

$$= M(I, \text{if is0?}(\text{push1}(x_1)) \text{ then push0}(\text{push1}(x_1)) \text{else push1}(\text{push1}(x_1)))$$

$$\begin{aligned} \text{is0?}^{\$}(M(I, \text{push1}(x_1))) &= \text{is0?}^{\$}(\text{push1}^{\$}(I(x_1))) \\ &= \text{is0?}^{\$}(\text{push1}^{\$}(011)) \\ &= \text{is0?}^{\$}(1011) \\ &= \text{FALSE} \end{aligned}$$

$$= M(I, \text{push1}(\text{push1}(x_1)))$$

$$= \text{push1}^{\$}(M(I, \text{push1}(x_1)))$$

$$= \text{push1}^{\$}(\text{push1}^{\$}(M(I, x_1)))$$

$$= \text{push1}^{\$}(\text{push1}^{\$}(I(x_1)))$$

$$= \text{push1}^{\$}(\text{push1}^{\$}(011))$$

$$= 11011$$

so 11011 is the answer.

[28]Strictly speaking, we must prove that the result is an expression. In order to do this we must give a formal definition of substitution. Let Exp_n be the set of expressions in which only $x_1, \ldots, x_n$ are the only individual variable symbols that appear, and let $u_1, \ldots, u_n$ be expressions. Define a function h: $\text{Exp}_n \to V^+$ as follows:

$$\begin{aligned} h(x_i) &= u_i, \qquad i = 1, \ldots, n \\ h(c) &= c \\ h(f(t_1, \ldots, t_n)) &= f(h(t_1), \ldots, h(t_n)) \\ h(\text{if } p(v_1, \ldots, v_n) \text{ then } t_1 \text{ else } t_2) &= \\ \text{if } p(h(v_1), \ldots, h(v_n)) &\text{ then } h(t_1) \text{ else } h(t_2) \\ h(F(t_1, \ldots, t_n)) &= F(h(t_1), \ldots, h(t_n)) \end{aligned}$$

Then $h(t)$ is the result of substituting u_i for each x_i in t. (This *defines* substitution.) By induction on t, one can now prove that for any expression t, $h(t)$ is also an expression.

In fact, it can be proven in our minilanguage that, if call-by-value gives an answer, then call-by-name gives the same answer. Since both calling methods seem equally plausible, this result sounds reasonable. In fact it is not: it depends crucially on our exclusion of side-effects. To see how side-effects destroy this equivalence, consider the following procedure written using the syntax of PASCAL:

```
procedure swap (x: real; y: real);
var t: real;
begin {snapshot taken here}
        t:=x;
        x:=y;
        y:=t
end
```

Imagine it is called from a main program via the call *swap*$(i,A[i])$. If call-by-value is used, then local copies of x and y are made, the procedure body manipulates the local copies, and the array A is unchanged [see Figure 3.5.1(a)}. If call by name is used, the effect is to execute

```
begin
        t:=i;
        i:=A[i];
    A[i]:=t
end
```

Unfortunately, in the third assignment statement, the subscript i has been changed, and the wrong element of the array gets updated [see Figure 3.5.1(b)]. If call-by-reference is used, then the swap takes place as desired [see Figure 3.5.1(c)].

In our restricted language, however, we do not have any of this pathological behavior, nor do we draw diagrams like Figure 3.5.1. Let us do one more example in the data type $\mathfrak{N}$ in our language of expressions.

EXAMPLE 3. Let

$$\delta\,\underline{F} = \underline{\text{if } x_1 > x_2 \text{ then } x_1 \text{ else } x_2}$$

$$I = \{(\underline{x}, 7), (\underline{y}, 9), (\underline{z}, 5)\}$$

$$t = \underline{F(F(\dot{-}(x,y), \dot{-}(y,z)), \dot{-}(x,z))}$$

Using call-by-value, we proceed as follows:

$$M(I, \underline{F(F(\dot{-}(x,y), \dot{-}(y,z)), \dot{-}(x,z))})$$

$$I'(\underline{x}_1) = M(I, \underline{F(\dot{-}(x,y), \dot{-}(y,z))})$$

$$I''(\underline{x}_1) = M(I, \underline{\dot{-}(x,y)}) = 0$$

$$I''(\underline{x}_2) = M(I, \underline{\dot{-}(y,z)}) = 4$$

$$= M(I'', \underline{\text{if } x_1 > x_2 \text{ then } x_1 \text{ else } x_2})$$

$$>^{\mathfrak{M}}(M(I'', \underline{x}_1), M(I'' \underline{x}_2)) = >^{\mathfrak{M}}(0,4) = \text{FALSE}$$

$$= M(I'', \underline{x}_2)$$

$$= 4$$

$$I'(\underline{x}_2) = M(I, \underline{\dot{-}(x,z)}) = 2$$

$$= M(I', \underline{\text{if } x_1 > x_2 \text{ then } x_1 \text{ else } x_2})$$

$$>^{\mathfrak{M}}(M(I', \underline{x}_1), M(I', \underline{x}_2)) = >^{\mathfrak{M}}(4,2) = \text{TRUE}$$

$$= M(I', \underline{x}_1)$$

$$= 4$$

This example points up several important characteristics of a call-by-value computation. At the first step, the expression $\underline{F(F(\dot{-}(x,y), \dot{-}(y,z)), \dot{-}(x,z))}$ was analyzed as a call on the function $\underline{F}$, with actual parameters $\underline{F(\dot{-}(x,y), \dot{-}(y,z))}$ and $\underline{\dot{-}(x,z)}$ (see Figure 3.5.2). We therefore began a subsidiary calculation to create an environment I' with $I'(\underline{x}_1) = M(I, \underline{F(\dot{-}(x,y), \dot{-}(y,z))})$ and $I'(\underline{x}_2) = M(I, \underline{\dot{-}(x,z)})$. The calculation of $I'(\underline{x}_1)$ was again a function call, so we started another subsidiary calculation. Note that in a subsidiary calculation, the actual parameters are evaluated in the *calling environment*. Thus, in the third line of the calculation, we wrote $M(I, \underline{\dot{-}(x,y)})$ rather than $M(I', \underline{\dot{-}(x,y)})$; the latter does not make sense for two reasons: first, I' is what we are trying to build, so we cannot use it yet; and second, I' will only really be defined on $\underline{x}_1$ and $\underline{x}_2$, so we cannot use it to evaluate $\underline{\dot{-}(x,y)}$.

This pattern of subsidiary calculations inside subsidiary calculations is characteristic of call-by-value computations. Note that we create new environments doing call-by-value, but we only use expressions that already exist in the problem.

By contrast, let us do the same example using call-by-name:

EXAMPLE 4

$$\delta\,\underline{F} = \underline{\text{if } x_1 > x_2 \text{ then } x_1 \text{ else } x_2}$$

$$I = \{(\underline{x}, 7), (\underline{y}, 9), (\underline{z}, 5)\}$$

$$t = \underline{F(F(\dot{-}(x,y), \dot{-}(y,z)), \dot{-}(x,z))}$$

Before call:

A[1]	A[2]	A[3]	i
4	3	5	2

At Snapshot:

x	y	t	A[1]	A[2]	A[3]	i
2	3		4	3	5	2

Before Return:

x	y	t	A[1]	A[2]	A[3]	i
3	2	2	4	3	5	2

(a)

At Snapshot:

t	A[1]	A[2]	A[3]	i
	4	3	5	2

Before Return:

t	A[1]	A[2]	A[3]	i
2	4	3	2	3

(b)

At Snapshot:

t	A[1]	y A[2]	A[3]	x i
	4	3	5	2

Before Return:

t	A[1]	y A[2]	A[3]	x i
2	4	2	5	3

(c)

FIGURE 3.5.1 A digression: procedure calling in a language with assignment. (a) call-by-value; (b) call-by-name; (c) call-by-reference.

$$M\big(I, \underline{\mathrm{F(F(\dot{-}(x,y), \dot{-}(y,z)), \dot{-}(x,z))}}\big)$$

$$= M\big(I, \underline{\text{if } \mathrm{F(\dot{-}(x,y), \dot{-}(y,z)) > \dot{-}(x,z)} \text{ then } \mathrm{F(\dot{-}(x,y), \dot{-}(y,z))} \text{ else } \mathrm{\dot{-}(x,z)}}\big)$$

$$>^{\mathfrak{N}}(M(I, \underline{\mathrm{F(\dot{-}(x,y), \dot{-}(y,z))}}\ ), M(I, \underline{\mathrm{\dot{-}(x,z)}})\ )$$

$$= >^{\mathfrak{N}}(M(I, \underline{\text{if } \mathrm{\dot{-}(x,y) > \dot{-}(y,z)} \text{ then } \mathrm{\dot{-}(x,y)} \text{ else } \mathrm{\dot{-}(y,z)}}\ ), 2)$$

$$>^{\mathfrak{N}}\big(M\big(I, \underline{\mathrm{\dot{-}(x,y)}}\ \big), M\big(I, \underline{\mathrm{\dot{-}(y,z)}}\ \big)\big)$$

$$= >^{\mathfrak{N}}(0,4)$$

$$= \mathrm{FALSE}$$

$$= >^{\mathfrak{N}}(M(I, \underline{\mathrm{\dot{-}(y,z)}}\ ), 2)$$

$$= >^{\mathfrak{N}}(4,2)$$

$$= \mathrm{TRUE}$$

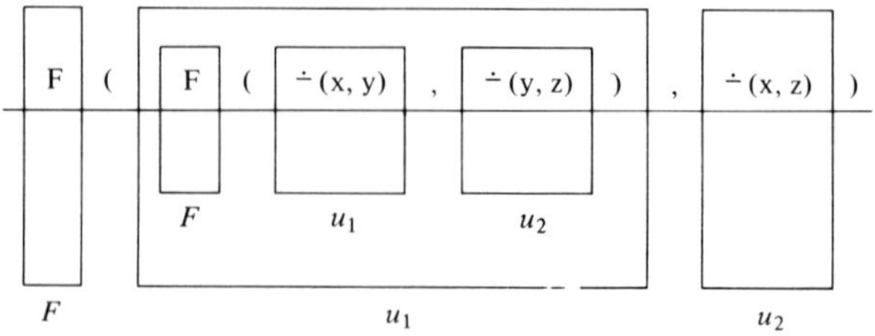

FIGURE 3.5.2 Parsing diagram for Example 3.

$$= M(I, \underline{\mathrm{F}(\dot{-}(\mathrm{x},\mathrm{y}), \dot{-}(\mathrm{y},\mathrm{z}))})$$

$$= M(I, \underline{\text{if } \dot{-}(\mathrm{x},\mathrm{y}) > \dot{-}(\mathrm{y},\mathrm{z}) \text{ then } \dot{-}(\mathrm{x},\mathrm{y}) \text{ else } \dot{-}(\mathrm{y},\mathrm{z})})$$

$$\boxed{\begin{aligned} &>^{\mathfrak{N}}(M(I, \underline{\dot{-}(\mathrm{x},\mathrm{y})}), M(I, \underline{\dot{-}(\mathrm{y},\mathrm{z})})) \\ &= >^{\mathfrak{N}}(0,4) \\ &= \mathrm{FALSE} \end{aligned}}$$

$$= M(I, \underline{\dot{-}(\mathrm{y},\mathrm{z})})$$

$$= 4$$

Again, in the first step we applied the call-by-name rule to the function $\underline{\mathrm{F}}$ with actual parameters $\underline{\mathrm{F}(\dot{-}(\mathrm{x},\mathrm{y}), \dot{-}(\mathrm{y},\mathrm{z}))}$ and $\underline{\dot{-}(\mathrm{x},\mathrm{z})}$. Aside from that, the call-by-name computation is quite different from the one for call-by-value. No new environments are created; the only subsidiary calculations are associated with conditionals.[29] On the other hand, new expressions may be created as in the second line of the calculation.

Another contrast between call-by-name and call-by-value is that in call-by-name, an actual parameter may be evaluated many times during the calculation [e.g., $M(I, \underline{\dot{-}(\mathrm{y},\mathrm{z})})$ is computed three times in our example], while in call-by-value, an actual parameter is evaluated exactly once, when the new environment is created, and is henceforth retrieved from the environment using the corresponding formal parameter. In other words, in call-by-value, the formal parameter is used as a name for the value of the actual parameter; in call-by-name, no such abbreviation is possible.[30]

[29]In fact, it is possible to define call-by-name so that there are no subsidiary calculations at all [Manna and Vuillemin 72].

[30]The use of a variable as a *name* for a quantity, so that one need not recompute the quantity, is a basic idea, both in mathematics and in programming; we use this idea again in Section 6.6.

EXERCISES 3.5

1. For each of the following strings, either show that the string is an expression or show that it is not an expression. Let r(F) = 1, r(G) = 2, r(H) = 2.

   (a) F(pop(y))
   (b) push0(if is0?(x) then x else y) [cf. Exercise 3.4.2]
   (c) if is1?(x) then H(pop(y)) else x
   (d) if is1?(pop(F(push1(y)))) then G(y, pop(x))else F(push0(x))

2. Let I = {(x, 00), (y, 111)}, J = {(x, 111), (y, 10)}.
   Let δF = if is1?(pop(x_1)) then x_1 else push0(x_1), r(F) = 1
   Evaluate, using both call-by-value and call-by-name:

   (a) F(x) in environment J
   (b) F(y) in environment I
   (c) F(pop(x)) in environment I
   (d) F(pop(x)) in environment J

3. Let δG = if is1?(x_1) then F(pop(x_2)) else F(pop(x_1)), r(G) = 2, and let F, I, and J be as in Exercise 3.5.2. Evaluate, using both call-by-value and call-by-name:

   (a) push1(G(pop(y), x)) in environment I
   (b) F(G(push1(x), x)) in environment J

4. Let

   δF = if is1?(x) then push1(x)
   else if is0?(x) then push0(x) else empty

   Evaluate, using call-by-value

   if is0?(F(x)) then F(push0(x)) else empty

   at I(x) = 01.

3.6 SOME INTRIGUING EXAMPLES

The FORTRAN language has the restriction that a subroutine may not call itself, either directly or through some chain of calls. In our language of expressions, there is no such restriction. This was intentional. Our evaluation function M does not need to worry about storage allocation, return addresses, or any of the other concerns that led to the restriction. All of this is avoided by the use of mathematical induction, so M is perfectly capable of handling self-referential calls. In fact, this is precisely the thing that causes our programs to become interesting.

EXAMPLE 1. Let

$$\delta F = \underline{\text{if empty?}(x_1)\text{ then }x_1}$$
$$\underline{\text{else if is0?}(x_1)\text{ then pop}(x_1)}$$
$$\underline{\text{else push1}(F(\text{pop}(x_1)))}$$

$$I(\underline{x}) = 10101$$

We evaluate $\underline{F(x)}$ by using call by value.

$$M(I, \underline{F(x)})$$

$$\boxed{I_1(\underline{x}_1) = M(I, \underline{x}) = 10101}$$

$$= M(I_1, \underline{\text{if empty?}(x_1)\text{ then }x_1}$$
$$\underline{\text{else if is0?}(x_1)\text{ then pop}(x_1)}$$
$$\underline{\text{else push1}(F(\text{pop}(x_1)))})$$

$$= M(I_1, \underline{\text{if is0?}(x_1)\text{ then pop}(x_1)\text{ else push1}(F(\text{pop}(x_1)))})$$

$$= M(I_1, \underline{\text{push1}(F(\text{pop}(x_1)))})$$

$$= \text{push1}^{\$}(M(I_1, \underline{F(\text{pop}(x_1))}))$$

$$\boxed{I_2(\underline{x}_1) = M(I_1, \underline{\text{pop}(x_1)}) = 0101}$$

$$= \text{push1}^{\$}(M(I_2, \underline{\text{if empty?}(x_1)\text{ then }x_1}$$
$$\underline{\text{else if is0?}(x_1)\text{ then pop}(x_1)}$$
$$\underline{\text{else push1}(F(\text{pop}(x_1)))}))$$

$$= \text{push1}^{\$}(M(I_2, \underline{\text{if is0?}(x_1)\text{ then pop}(x_1)\text{ else push1}(F(\text{pop}(x_1)))}))$$

$$= \text{push1}^{\$}(M(I_2, \underline{\text{pop}(x_1)}))$$

$$= \text{push1}^{\$}(101)$$

$$= 1101$$

$M(I, \underline{F(x)})$ is $I(\underline{x})$ with the topmost 0 deleted; here we have omitted the subsidiary calculations for the predicates. Notice that $\underline{F}$ is defined in terms of itself: To evaluate $\underline{F(x)}$ you wind up evaluating $\underline{F(\text{pop}(x))}$ and doing a push1$^{\mathbb{S}}$ on the result. A function that calls itself is said to be *recursive*. This should not come as a great shock to you—we have been defining functions in terms of themselves throughout this book. This is what mathematical induction is about. And this mysterious thing called "recursion" is nothing more than induction applied to programming.

Let us do another example.

EXAMPLE 2. In the structure $\mathfrak{N}$ of nonnegative integers, let

$$\delta\,\underline{\mathrm{F}} = \underline{\mathrm{if}\ =0?(\mathrm{x}_1)\ \mathrm{then}\ 0\ \mathrm{else}\ +(\mathrm{x}_2, \mathrm{F}(\dot{-}(\mathrm{x}_1,1),\mathrm{x}_2))}$$

$$I(\underline{\mathrm{x}})=2$$

$$I(\underline{\mathrm{y}})=3$$

$$M(I,\ \underline{\mathrm{F}(\mathrm{x},\mathrm{y})}\ )$$

$$\boxed{\begin{aligned} I_1(\underline{\mathrm{x}}_1) &= M(I,\underline{\mathrm{x}})=2\\ I_1(\underline{\mathrm{x}}_2) &= M(I,\underline{\mathrm{y}})=3\end{aligned}}$$

$$= M(I_1, \underline{\mathrm{if}\ =0?(\mathrm{x}_1)\ \mathrm{then}\ 0\ \mathrm{else}\ +(\mathrm{x}_2,\mathrm{F}(\dot{-}(\mathrm{x}_1,1),\mathrm{x}_2))}\ )$$

$$= M(I_1, \underline{+(\mathrm{x}_2,\mathrm{F}(\dot{-}(\mathrm{x}_1,1),\mathrm{x}_2))}\ )$$

$$= +^{\mathfrak{N}}(M(I_1,\underline{\mathrm{x}}_2), M(I_1, \underline{\mathrm{F}(\dot{-}(\mathrm{x}_1,1),\mathrm{x}_2)}))$$

$$= +^{\mathfrak{N}}(3, M(I_1, \underline{\mathrm{F}(\dot{-}(\mathrm{x}_1,1),\mathrm{x}_2)}\ ))$$

$$\boxed{\begin{aligned} I_2(\underline{\mathrm{x}}_1) &= M(I_1, \underline{\dot{-}(\mathrm{x}_1,1)}\ )=1\\ I_2(\underline{\mathrm{x}}_2) &= M(I_1,\underline{\mathrm{x}}_2)=3\end{aligned}}$$

$$= +^{\mathfrak{N}}(3, M(I_2, \underline{\mathrm{if}\ =0?(\mathrm{x}_1)\ \mathrm{then}\ 0\ \mathrm{else}\ +(\mathrm{x}_2,\mathrm{F}(\dot{-}(\mathrm{x}_1,1),\mathrm{x}_2))}\ ))$$

$$= +^{\mathfrak{N}}(3, M(I_2, \underline{+(\mathrm{x}_2,\mathrm{F}(\dot{-}(\mathrm{x}_1,1),\mathrm{x}_2))}\ ))$$

$$= +^{\mathfrak{N}}(3, +^{\mathfrak{N}}(M(I_2,\underline{\mathrm{x}}_2), M(I_2, \underline{\mathrm{F}(\dot{-}(\mathrm{x}_1,1),\mathrm{x}_2)}\ )))$$

$$= +^{\mathfrak{N}}(3, M(I_2, \underline{\mathrm{F}(\dot{-}(\mathrm{x}_1,1),\mathrm{x}_2)}\ )))$$

$$\boxed{\begin{aligned} I_3(\underline{\mathrm{x}}_1) &= M(I_2, \underline{\dot{-}(\mathrm{x}_1,1)}\ )=0\\ I_3(\underline{\mathrm{x}}_2) &= M(I_2,\underline{\mathrm{x}}_2)=3\end{aligned}}$$

$= +^{\mathfrak{N}}(3, +^{\mathfrak{N}}(3, M(I_3, \underline{\text{if } =0?(x_1) \text{ then } 0 \text{ else } +(x_2, F(\dot{-}(x_1,1), x_2))})))$

$= +^{\mathfrak{N}}(3, +^{\mathfrak{N}}(3,0))$

$=6$

We shall prove in Section 8 that in any environment I, $M(I, \underline{F(x,y)}) = I(\underline{x}) \times I(\underline{y})$. The proof, needless to say, is by induction, in this case on the value of $I(\underline{x})$.

Recursion lets us do interesting things because it allows a computation to "run on." For terms and conditionals, computations are bounded: that is, for each conditional t, there is a number k such that, no matter what environment I we choose, the calculation of $M(I,t)$ never takes more than k steps. You can only do a certain number of tests before you get to a term. Once recursion is introduced, this restriction is lifted. In Example 1, we checked each element of the stack until we found a 0. This process could have continued as long as necessary, with new environments created, with each new value of $\underline{x}_1$ one shorter than the previous one, until either $\underline{x}_1$ starts with a zero or is empty.

Let us try another example: Let

$$\delta \underline{F} = \underline{F(x_1)}$$

$$I = \{(\underline{x}_1, 1)\}$$

$$M(I, \underline{F(x_1)})$$

$$\boxed{I'(\underline{x}_1) = 1}$$

$$= M(I', \underline{F(x_1)})$$

$$\boxed{I''(\underline{x}_1) = 1}$$

$$= M(I'', F(\underline{x}_1))$$

What has gone wrong here? Clearly, applying the call-by-value rule is never going to terminate. Unlike the previous examples, our environments are not changing. The call-by-name rule does not give termination either (try it and see!). What we have written is just 10 GO TO 10—an infinite loop. Just as FORTRAN programs often contain bugs that result in infinite loops, expressions may get into infinite loops in subtle ways (see Exercise 6.4).

But what is the value of $M(I, \underline{F(x_1)})$? M was supposed to be a function, was it not? In fact, M is only a *partial* function. Remember that a functional form definition, as introduced in Section 2.4, is only an abbreviation for a relational form definition. In this case, we are defining M, a subset of $\mathrm{FENV} \times \mathrm{ENV} \times \mathrm{Exp} \times A$. Since our definition of M was unambiguous, M is a partial function. But there is no $a \in A$ such that

$$\left(\delta, I, \underline{\mathrm{F}(\mathrm{x}_1)}, a\right)$$

is in M. So $M(I, \underline{\mathrm{F}(\mathrm{x}_1)})$ is indeed undefined.

EXERCISES 3.6

1. Let

$$\delta\underline{\mathrm{F}} = \underline{\text{if empty?}(\mathrm{x}_1) \text{ then empty}}$$
$$\underline{\text{else if is0?}(\mathrm{x}_1) \text{ then } \mathrm{x}_1}$$
$$\underline{\text{else F(pop}(\mathrm{x}_1))}$$

Evaluate $\underline{\mathrm{F}(\mathrm{x})}$ at $I(\underline{\mathrm{x}}) = 11010$

(a) using call-by-name
(b) using call-by-value

2. Let

$$\delta\underline{\mathrm{F}} = \underline{\text{if } \mathrm{x}_1 > 100 \text{ then } \mathrm{x}_1 - 10 \text{ else } \mathrm{F}(\mathrm{F}(\mathrm{x}_1 + 11))}\,.$$

Evaluate $\underline{\mathrm{F}(99)}$ in $\mathfrak{N}$, at any I.

(a) using call-by-name
(b) using call-by-value

3. Write a program in the language of stacks that removes all the 0's from a stack.

4. Let

$$\delta\underline{\mathrm{F}} = \underline{\text{if empty?}(\mathrm{x}_1) \text{ then } \mathrm{x}_1 \text{ else } \mathrm{F}(\text{pop}(\mathrm{x}_1), \mathrm{G}(\mathrm{x}_1))}$$

$$\delta\underline{\mathrm{G}} = \underline{\mathrm{G}(\text{push1}(\mathrm{x}_1))}$$

$$I(\underline{\mathrm{x}}) = 10$$

Evaluate $\underline{\mathrm{F}(\mathrm{x},\mathrm{x})}$ at I

(a) using call-by-name
(b) using call-by-value

(*Hint:* This is not an exercise in deforestation.)

3.7 PROGRAMMING BY INDUCTION

At this point it may be wise to look back on this chapter and say, "So what?" We have, using mathematical induction, defined a programming language and defined its computations. What has this gained us? The answer is that it has given us understanding:

It helps us understand how we can go about defining advanced programming languages.

It points up implementation decisions (e.g., call-by-name versus call-by-value).

Because we have a mathematical notion of what a program does, we can *prove* the correctness of programs.

Our understanding helps us to *write* programs.

It is this last point that we shall discuss in this section.

Consider the following facts about the set S of stacks:

(i) $\text{empty}^{\mathcal{S}} \in S$
(ii) if $z \in S$, then $\text{push0}^{\mathcal{S}}(z) \in S$ and $\text{push1}^{\mathcal{S}}(z) \in S$.

But every stack is obtained by pushing 0's and 1's on the empty stack. So we may add another fact about S:

(iii) nothing else

This is an inductive definition of the set of stacks. Hence, if we want to define a function on the set of stacks we should do it inductively as well.

EXAMPLE 1. Let us define a function f: $S \to S$ that changes every 0 on the stack to a 1 and vice versa.

	S	f
(i)	$\text{empty}^{\mathcal{S}} \in S$	$f(\text{empty}^{\mathcal{S}}) = \text{empty}^{\mathcal{S}}$
(ii)	$\text{push0}^{\mathcal{S}}(z) \in S$	$f(\text{push0}^{\mathcal{S}}(z)) = \text{push1}^{\mathcal{S}}(f(z))$
(iii)	$\text{push1}^{\mathcal{S}}(z) \in S$	$f(\text{push1}^{\mathcal{S}}(z)) = \text{push0}^{\mathcal{S}}(f(z))$

If x is the empty stack, then $f(x)$ is empty; if x has a 0 on top, then $x = \text{push0}^{\mathcal{S}}(z)$ and $f(x)$ is the stack obtained by taking f of z and pushing a 1 onto that; if $x = \text{push1}^{\mathcal{S}}(z)$, then $f(x)$ is the stack obtained by pushing a 0 onto $f(z)$.

This should look suspiciously like a conditional expression. Recalling that if x is $\text{push0}^{\mathcal{S}}(z)$ or $\text{push1}^{\mathcal{S}}(z)$, then $z = \text{pop}^{\mathcal{S}}(x)$, we write

$$\delta\,\underline{\mathrm{F}} = \underline{\text{if empty?}(\mathrm{x}_1)\text{ then empty}}$$
$$\underline{\text{else if is0?}(\mathrm{x}_1)\text{ then push1}(\mathrm{F}(\text{pop}(\mathrm{x}_1)))}$$
$$\underline{\text{else push0}(\mathrm{F}(\text{pop}(\mathrm{x}_1)))}$$

[We do not need an is1?(x_1) on the last line since if neither empty?(x_1) nor is0?(x_1) is true, then is1?(x_1) is true.]

This function exchanges 0's and 1's on its argument.

EXAMPLE 2 ("Append"). We want to define a function f of two stacks that has as its value the second stack with the first stack copied on top of it. Thus $f(01,000)=01000$. Consider $f(x,y)$. If x is the empty stack, then $f(x,y)=y$. If x starts with a 0 (or 1), then $f(x,y)$ is obtained by copying the rest of x onto y, and then pushing a 0 (or 1) onto the result. More concisely,

$$\begin{aligned}
&f(\mathrm{empty}^{\mathbb{S}},y)=y\\
&f(\mathrm{push0}^{\mathbb{S}}(x),y)=\mathrm{push0}^{\mathbb{S}}(f(x,y))\\
&f(\mathrm{push1}^{\mathbb{S}}(x),y)=\mathrm{push1}^{\mathbb{S}}(f(x,y))
\end{aligned}$$

or

$$\begin{aligned}
\delta\,\underline{\mathrm{F}} = {}&\underline{\text{if empty?}(\mathrm{x}_1)\text{ then }\mathrm{x}_2}\\
&\quad\underline{\text{else if is0?}(\mathrm{x}_1)\text{ then push0}(\mathrm{F}(\mathrm{pop}(\mathrm{x}_1),\mathrm{x}_2))}\\
&\quad\underline{\text{else push1}(\mathrm{F}(\mathrm{pop}(\mathrm{x}_1),\mathrm{x}_2))}
\end{aligned}$$

Letting $I(\underline{\mathrm{x}})=01$ and $I(\underline{\mathrm{y}})=000$, and using call by name,

$$\begin{aligned}
M(I,\ \underline{\mathrm{F}(\mathrm{x},\mathrm{y})}\ ) &= M(I,\ \underline{\text{if empty?}(\mathrm{x})\text{ then y}}\\
&\qquad\underline{\text{else if is0?}(\mathrm{x})\text{ then push0}(\mathrm{F}(\mathrm{pop}(\mathrm{x}),\mathrm{y}))}\\
&\qquad\underline{\text{else push1}(\mathrm{F}(\mathrm{pop}(\mathrm{x}),\mathrm{y}))}\ )\\
&= M(I,\ \underline{\text{if is0?}(\mathrm{x})\text{ then push0}(\mathrm{F}(\mathrm{pop}(\mathrm{x}),\mathrm{y}))}\\
&\qquad\underline{\text{else push1}(\mathrm{F}(\mathrm{pop}(\mathrm{x}),\mathrm{y}))}\ )\\
&= M(I,\ \underline{\text{push0}(\mathrm{F}(\mathrm{pop}(\mathrm{x}),\mathrm{y}))}\ )\\
&= \mathrm{push0}^{\mathbb{S}}(M(I,\ \underline{\mathrm{F}(\mathrm{pop}(\mathrm{x}),\mathrm{y})}\ ))\\
&= \mathrm{push0}^{\mathbb{S}}(M(I,\ \underline{\text{if empty?}(\mathrm{pop}(\mathrm{x}))\text{ then y}}\\
&\qquad\underline{\text{else if is0?}(\mathrm{pop}(\mathrm{x}))}\\
&\qquad\quad\underline{\text{then push0}(\mathrm{F}(\mathrm{pop}(\mathrm{pop}(\mathrm{x})),\mathrm{y}))}\\
&\qquad\quad\underline{\text{else push1}(\mathrm{F}(\mathrm{pop}(\mathrm{pop}(\mathrm{x})),\mathrm{y}))}\ ))\\
&= \mathrm{push0}^{\mathbb{S}}(M(I,\ \underline{\text{push1}(\mathrm{F}(\mathrm{pop}(\mathrm{pop}(\mathrm{x})),\mathrm{y}))}\ ))\\
&= \mathrm{push0}^{\mathbb{S}}(\mathrm{push1}^{\mathbb{S}}(M(I,\ \underline{\mathrm{F}(\mathrm{pop}(\mathrm{pop}(\mathrm{x})),\mathrm{y})}\ )))\\
&= \mathrm{push0}^{\mathbb{S}}(\mathrm{push1}^{\mathbb{S}}(M(I,\ \underline{\text{if empty?}(\mathrm{pop}(\mathrm{pop}(\mathrm{x}))\text{ then y}}\\
&\qquad\underline{\text{else}}\ \ldots)))\\
&= \mathrm{push0}^{\mathbb{S}}(\mathrm{push1}^{\mathbb{S}}(M(I,\ \underline{\mathrm{y}}\,)))\\
&= \mathrm{push0}^{\mathbb{S}}(\mathrm{push1}^{\mathbb{S}}(I(\,\underline{\mathrm{y}}\,)))\\
&= \mathrm{push0}^{\mathbb{S}}(\mathrm{push1}^{\mathbb{S}}(000))=01000
\end{aligned}$$

which is the desired answer.[31]

[31]Computers are much better at this than people are.

EXAMPLE 3 (Substitution). g is to be a two-place function; $g(x,y)$ will substitute a copy of x for each "1" in y:

if y is empty$^{\mathbb{S}}$, then $g(x,y) = \text{empty}^{\mathbb{S}}$
if $y = \text{push0}^{\mathbb{S}}(z)$, then $g(x,y) = \text{push0}^{\mathbb{S}}(g(x,z))$ ("copy the 0 and go on")
if $y = \text{push1}^{\mathbb{S}}(z)$, then $g(x,y)$ is obtained by copying x onto the top of thing obtained by substituting x for 1 in z: $f(x,g(x,z))$, using the f of the previous example. (Here f is called a "help function.")

So

$$\delta\,\underline{G} = \underline{\text{if empty?}(x_2)\text{ then empty}}$$
$$\underline{\text{else if is0?}(x_2)\text{ then push0}(G(x_1,\text{pop}(x_2)))}$$
$$\underline{\text{else }F(x_1,G(x_1,\text{pop}(x_2)))}$$

Moral: When a set is defined by induction, the best way to define a function on it is by induction; the best way to program the function is by recursion.

EXERCISES 3.7

Write functions that do the following to a stack or stacks: use help functions where necessary.

1. Change every 1 to a 0.

2. Delete every 0.

3. Change every 0 to a 1 and every (original) 1 to a 00.

4. Given two stacks, return a stack consisting of as many 1's as there are in both stacks combined.

5. Change every occurrence of the pattern 01 to 10.

6. Return a stack half the length of the original (you may round off $2\frac{1}{2}$ to 2).

7. A while-loop program

```
begin
    x2 ← a;
    while B do begin
        x2 ← g(x1, x2);
        x1 ← f(x1)
        end
end
```

leaving the answer in x_2, may be replaced by a function

$$\underline{\delta}\underline{F} = \underline{LOOP}(x_1, a)$$
$$\underline{\delta LOOP} = \underline{if}\ B\ \underline{then\ LOOP(f(x_1), g(x_1, x_2))}$$
$$\underline{else\ x_2}$$

Use this idea to write a recursive multiplication function based on the program of Figure 2.1.1.

3.8 PROVING RECURSIVE PROGRAMS CORRECT

In this section we shall prove the correctness of programs in the language of expressions. As usual, our proofs will be by induction.

Theorem 3.8.1. *Let*

$$\delta\,\underline{F} = \underline{\text{if empty?}(x_1)\text{ then empty}}$$
$$\underline{\text{else if is0?}(x_1)\text{ then push1}(F(\text{pop}(x_1)))}$$
$$\underline{\text{else push0}(F(\text{pop}(x_1)))}$$

Then in the data type $\mathbb{S}$, *for every interpretation* I, *the value of* $M(I, \underline{F(x_1)})$ *(using call by value) is the stack obtained by changing every* 1 *to a* 0 *and vice versa.*

PROOF. The induction hypothesis is, "If $I(\underline{x}_1)$ is a stack of length $\leqslant k$, then $M(\delta, I, \delta\underline{F})$ is defined and equals the stack obtained from $I(\underline{x}_1)$ by changing 0's to 1's and vice versa." Call this stack the *complement* of the original.

Base Step: ($k=0$): If $I(\underline{x}_1) = \text{empty}^{\mathbb{S}}$, then $M(I, \delta\underline{F}) = \text{empty}^{\mathbb{S}}$, which is the right answer.

Induction Step: There are two cases:

Case I:

$$I(\underline{x}_1) = \text{push0}^{\mathbb{S}}(y).$$

By IH, if $I_2(\underline{x}_1) = y$, then $M(I_2, \delta\underline{F})$ gives the result of exchanging 0's and 1's in y, so the right answer is $\text{push1}^{\mathbb{S}}(M(I_2, \delta\underline{F}))$.

$$M(I, \underline{F(x_1)})$$

$$\boxed{I_1(\underline{x}_1) = M(I, \underline{x}_1) = \text{push0}^{\mathbb{S}}(y)}$$

$$= M(I_1,\ \underline{\text{if empty?}(x_1)\text{ then empty}}$$
$$\underline{\text{else if is0?}(x_1)\text{ then push1}(F(\text{pop}(x_1)))}$$
$$\underline{\text{else push0}(F(\text{pop}(x_1)))}\ )$$

$$= M(I_1, \underline{\text{push1}(\text{F}(\text{pop}(\text{x}_1)))})$$

$$= \text{push1}^{\mathbb{S}}(M(I_1, \underline{\text{F}(\text{pop}(\text{x}_1))}))$$

$$\boxed{\begin{aligned} I_2(\underline{\text{x}}_1) &= M(I_1, \underline{\text{pop}(\text{x}_1)}) \\ &= \text{pop}^{\mathbb{S}}(M(I_1, \underline{\text{x}}_1)) \\ &= \text{pop}^{\mathbb{S}}(\text{push0}^{\mathbb{S}}(y)) \\ &= y \end{aligned}}$$

$$= \text{push1}^{\mathbb{S}}(M(I_2, \delta\underline{\text{F}}))$$

Case II:

$$I(\underline{\text{x}}_1) = \text{push1}^{\mathbb{S}}(\text{y})$$

Similar.

So for every I', $M(I',\delta F)$ gives the complement of $I'(\underline{\text{x}_1})$. So for every I, $M(I,\underline{\text{F}(\text{x})}) = M(I',\delta F)$ = the complement of $I'(\underline{\text{x}_1})$ = the complement of $I(\underline{\text{x}_1})$. □

Unfortunately, this proof involves reasoning about the phrase "changing all 0's into 1's and vice versa." So long as we rely on words to express the desired effect of a program (and this is typically the case), then some reasoning about words will be necessary. However, the difficult part of the proof—dealing with the *dynamic* aspects of the program, such as the pattern of recursive calls—is handled absolutely rigorously by our meaning function M and by induction.[32]

On the other hand, one is often lucky enough so that the desired result can be expressed mathematically:

Theorem 3.8.2. *Let* $\delta\underline{\text{F}} = \underline{\text{if } {=}0?(\text{x}_1) \text{ then } 0 \text{ else } +(\text{x}_2, \text{F}(\dot{-}(\text{x}_1, 1), \text{x}_2))}$, *and let M use call-by-value. Then in the data type $\mathfrak{N}$, for any environment I,* $M(I, \underline{\text{F}(\text{x},\text{y})}) = I(\underline{\text{x}}) \times I(\underline{\text{y}})$.

PROOF. Again we proceed by induction on k, with the following induction hypothesis: If $I(\underline{\text{x}_1}) = k$, then $M(I, \delta\underline{\text{F}}) = I(\underline{\text{x}_1}) \times I(\underline{\text{x}_2})$. If this is true for all k, then the conclusion of the theorem follows by a single application of the call-by-value rule.

[32]See, for example, how the confusion between the name of a variable and its value at different times in the program, which was a source of confusion in the proof of Theorem 2.1.1 (see footnote 2.1), is eliminated by the explicit use of environments.

The technique of proof we are using is similar to *subgoal induction* [Morris and Wegbreit 77]. For a discussion of related methods (especially for use with call-by-name), see [Manna and Vuillemin 72].

Base Step: $(k=0)$: If $I(\underline{\mathrm{x}}_1)=0$, then $M(I,\delta\underline{\mathrm{F}})=0$, by a simple calculation.

Induction Step: Assume that if $I'(\underline{\mathrm{x}}_1)=k$, then $M(I',\delta\underline{\mathrm{F}})=I'(\underline{\mathrm{x}}_1)\times I'(\underline{\mathrm{x}}_2)$. Let $I(\underline{\mathrm{x}}_1)=k+1$. Then

$$M(I,\ \underline{\text{if } =0?(\mathrm{x}_1) \text{ then } 0 \text{ else } +(\mathrm{x}_2, \mathrm{F}(\dot{-}(\mathrm{x}_1,1),\mathrm{x}_2))}\ )$$

$$= M(I,\ \underline{+(\mathrm{x}_2,\mathrm{F}(\dot{-}(\mathrm{x}_1,1),\mathrm{x}_2))}\ )$$

$$= I(\underline{\mathrm{x}}_2) + M(I,\ \underline{\mathrm{F}(\dot{-}(\mathrm{x}_1,1),\mathrm{x}_2)}\ )$$

$$\boxed{\begin{aligned} &I'(\underline{\mathrm{x}}_1) = M(I,\ \underline{\dot{-}(\mathrm{x}_1,1)}\ ) = \dot{-}^{\mathfrak{N}}(M(I,\underline{\mathrm{x}}_1), M(I,\underline{1}))\\ &\quad = \dot{-}^{\mathfrak{N}}(k+1,1)=k\\ &I'(\underline{\mathrm{x}}_2) = I(\underline{\mathrm{x}}_2)\end{aligned}}$$

$$= I(\underline{\mathrm{x}}_2) + M(I',\delta\,\underline{\mathrm{F}})$$

$$= I(\underline{\mathrm{x}}_2) + (k\times I(\underline{\mathrm{x}}_2)) \qquad \text{(by IH)}$$

$$= (k+1)\times I(\underline{\mathrm{x}}_2) \qquad \text{(by the laws of arithmetic)}$$

$$= I(\underline{\mathrm{x}}_1)\times I(\underline{\mathrm{x}}_2).$$

□

EXERCISES 3.8

Use call by value for these exercises. All use the data type $\mathfrak{N}$.

1. Let

$$\delta\,\underline{\mathrm{F}} = \underline{\text{if } =0?(\mathrm{x}_2) \text{ then } \mathrm{x}_1}$$
$$\underline{\text{else if } =0?(\mathrm{x}_1) \text{ then } 0}$$
$$\underline{\text{else } \mathrm{F}(\dot{-}(\mathrm{x}_1,1), \dot{-}(\mathrm{x}_2,1))}$$

Prove $M(I,\delta\underline{\mathrm{F}}) = I(\underline{\mathrm{x}}_1)\dot{-}I(\underline{\mathrm{x}}_2)$.

2. Let

$$\delta\,\underline{\mathrm{F}} = \underline{\text{if } =0?(\mathrm{x}_2) \text{ then } \mathrm{x}_1}$$
$$\underline{\text{else if } =0?(\mathrm{x}_1) \text{ then } \mathrm{x}_2}$$
$$\underline{\text{else } \mathrm{F}(\dot{-}(\mathrm{x}_1,1), \dot{-}(\mathrm{x}_2,1))}$$

Prove

$$M(I,\delta \underline{F}) = \begin{cases} I(\underline{x}_1) - I(\underline{x}_2) & \text{if } I(\underline{x}_1) \geq I(\underline{x}_2) \\ I(\underline{x}_2) - I(\underline{x}_1) & \text{if } I(\underline{x}_1) < I(\underline{x}_2) \end{cases}$$

3. Let

$$\delta \underline{F} = \underline{\text{if } x_2 > x_1 \text{ then } x_1 \text{ else } +(1, F(\dot{-}(x_1, x_2), x_2))}$$

Prove $M(I, \delta \underline{F}) = I(\underline{x}_1)$ div $I(\underline{x}_2)$ (integer quotient).

(*Hint*: Use "if $I(\underline{x}_1) \leq k \ldots$" in the induction hypothesis.)

4 Programs That Manipulate Programs

Up to now, our programs have manipulated fairly simple kinds of data, such as integers or stacks of 0's and 1's. In the real world, however, programs manipulate much more complex kinds of data. In this chapter we shall see what happens when we try to manipulate *programs* as data. Compilers, interpreters, and operating systems are typical programs that operate on other programs. We shall see that in our language of expressions it is no harder to manipulate programs than any other kind of data.

4.1 THE DATA TYPE OF LISTS

If we are going to manipulate programs, we must have a data type in which programs can conveniently be expressed. Clearly, integers and stacks are inappropriate, and strings turn out to be not much better. The data type we shall use is the data type of *lists*.

Let At be a fixed set whose members are called *atoms*. The set L of lists over At is the subset of $(At \cup \{(,)\})^+$ defined by

(i) if $a \in At$, then a is a list
(ii) () is a list
(iii) if $l_1, \ldots, l_n$ are lists, then $(l_1 \ldots l_n)$ is a list
(iv) nothing else

Thus, if X, FOO, BAR, and BAZ are atoms, typical lists are

FOO
()
(X FOO BAZ) (a list with three elements)
((X FOO) BAR) (a list with two elements)
(X X X X X) (a list with five elements, each of which is the atom X)

We shall typically separate adjacent atoms by spaces. In keeping with our practice of underlining strings of formal symbols, we use "squiggly underline" ( ) beneath the atoms and parentheses. Typical atoms are those in the preceding example; we shall occasionally assume every nonnegative integer is an atom.

If it is possible to distinguish adjacent atoms, then this turns out to be an unambiguous definition, so we can define functions on L by induction.

There are three basic functions on L. In these definitions a denotes an atom, and $l_1, \ldots, l_n$ denote lists.

1. $\text{first}^{\mathcal{L}}: L \rightarrow L$
   $\text{first}^{\mathcal{L}}(a) = ()$
   $\text{first}^{\mathcal{L}}(\,()\,) = ()$
   $\text{first}^{\mathcal{L}}(\,(l_1 \ldots l_n)\,) = l_1$
2. $\text{rest}^{\mathcal{L}}: L \rightarrow L$
   $\text{rest}^{\mathcal{L}}(a) = ()$
   $\text{rest}^{\mathcal{L}}(\,()\,) = ()$
   $\text{rest}^{\mathcal{L}}(\,(l_1 \ldots l_n)\,) = (l_2 \ldots l_n)$
3. $\text{build}^{\mathcal{L}}: L \times L \rightarrow L$
   $\text{build}^{\mathcal{L}}(l, a) = a$
   $\text{build}^{\mathcal{L}}(l, ()\,) = (l)$
   $\text{build}^{\mathcal{L}}(l, (l_1 \ldots l_n)\,) = (l\ l_1 \ldots l_n)$

These functions are similar, but not identical, to ones in LISP.[1]

We need some predicates; we choose two:

$$\text{atom?}^{\mathcal{L}}(x) = \begin{cases} \text{TRUE} & \text{if } x \in At \\ \text{FALSE} & \text{otherwise} \end{cases}$$

$$\text{eq?}^{\mathcal{L}}(x,y) = \begin{cases} \text{TRUE} & \text{if } x = y \\ \text{FALSE} & \text{otherwise} \end{cases}$$

We shall have a constant for each atom. Constant symbols will be underlined in the usual way; hence FOO is a constant symbol denoting the atom FOO. Among the atoms are true and false. We also include the constant symbol nil, denoting (), called the null list.

To summarize, we have in our data type $\mathcal{L}$

the universe $\mathcal{L}$,

the functions $\text{first}^{\mathcal{L}}$, $\text{rest}^{\mathcal{L}}$, $\text{build}^{\mathcal{L}}$,

the predicates $\text{atom?}^{\mathcal{L}}$, $\text{eq?}^{\mathcal{L}}$,

the constant $\text{nil}^{\mathcal{L}}$ and a constant for every atom.

[1]Note that in LISP, nil is an atom, whereas it is not in $\mathcal{L}$. Our first, rest, and build correspond to LISP's car, cdr, and cons, except that car(a) and cdr(a) are undefined when a is an atom, and cons(x,a) is not a list when a is a nonnull atom (its value is then something called a "dotted pair"). Our eq is LISP's "equal."

We can now write some useful auxiliary functions to take the second element, third element, etc., of a list, and to build lists of length 2, 3, and so on:

$\delta\underline{\mathrm{2ND}} = \underline{\mathrm{first(rest}(x_1))}$

$\delta\underline{\mathrm{3RD}} = \underline{\mathrm{2ND(rest}(x_1))}$

$\delta\underline{\mathrm{4TH}} = \underline{\mathrm{3RD(rest}(x_1))}$

$\delta\underline{\mathrm{5TH}} = \underline{\mathrm{4TH(rest}(x_1))}$

$\delta\underline{\mathrm{2LIST}} = \underline{\mathrm{build}(x_1, \mathrm{build}(x_2, \mathrm{nil}))}$

$\delta\underline{\mathrm{3LIST}} = \underline{\mathrm{build}(x_1, \mathrm{2LIST}(x_2, x_3))}$

Let us assume that the integers are available as atoms, and that the usual operations in $\mathfrak{N}$ are available also.[2] Then we can write a function that, given a list $\underline{x}_1$ and a nonnegative integer $\underline{x}_2 \geqslant 1$, finds the $\underline{x}_2$th element of $\underline{x}_1$, or returns the atom error if there is no such element (because the list is too short).

$$\delta\,\underline{\mathrm{NTH}} = \underline{\text{if eq?}(x_1, \text{nil}) \text{ then error}}$$
$$\underline{\text{else if } =(x_2, 1) \text{ then first}(x_1)}$$
$$\underline{\text{else NTH(rest}(x_1), \dot{-}(x_2, 1))}$$

If the value of $\underline{x}_1$ is the null list, then it has no elements. The first element of a list may be obtained by taking first$^{\mathfrak{L}}$ of it; otherwise we ask for the $(\underline{x}_2 - 1)$th element of $\underline{\text{rest}(x_1)}$.[3]

If the language of expressions is to be a convenient programming language, it is going to need mnemonic variable names. We therefore introduce them, resulting in the following definition for $\underline{\mathrm{NTH}}$:

$$\delta\,\underline{\mathrm{NTH}(x, n)} = \underline{\text{if eq?}(x, \text{nil}) \text{ then error}}$$
$$\underline{\text{else if } =(n, 1) \text{ then first}(x)}$$
$$\underline{\text{else NTH(rest}(x), \dot{-}(n, 1))}$$

Such mnemonic formal parameters should be regarded as $\underline{x}_1$, $\underline{x}_2$, etc., written down in a special purpose alphabet.

Let us try writing some more useful functions. We can represent finite sets (of, say, integers) by representing

$$\{n_1, n_2, \ldots, n_k\}$$

as

$$(n_1\ n_2\ \ldots\ n_k)$$

[2]We can assume these give nonsense values if given a noninteger as an argument.

[3]Notice how the confusion between name and variable has crept back in! It seems quite difficult to avoid such confusion when talking informally about programs; this is another good reason for doing formal proofs of correctness.

i.e., as a list of integers without repetitions.[4] We can then write a function that takes such a list s and an integer n, and returns either true$^{\mathfrak{L}}$ or false$^{\mathfrak{L}}$, depending on whether n is a member of s:[5]

$$\delta\ \underline{\text{MEMBER(n, s)}} = \underline{\text{if eq?(s, nil) then false}}$$
$$\underline{\text{else if eq?(first(s), n) then true}}$$
$$\underline{\text{else MEMBER(n, rest(s))}}$$

If s is the empty list, then it has no members, so the answer must be false; if the first member of the list is the sought-for integer n, then the answer is true; otherwise, we look for n in the rest of the list. Each time we recurse, the value of s gets shorter, so eventually we shall get to an answer, either by finding the integer or by having the value of s reach nil.

Another useful function on finite sets of integers is REMBER, which takes a set S and an integer n and computes $S-\{n\}$, that is, S with the element n removed (if it is present):

$$\delta\ \underline{\text{REMBER(n, s)}} = \underline{\text{if eq?(s, nil) then nil}}$$
$$\underline{\text{else if eq?(first(s), n) then rest(s)}}$$
$$\underline{\text{else build(first(s),REMBER(n, rest(s)))}}$$

If the list denoted by s is empty, then subtracting an element from it gives empty. If its first element is the one we are looking for, then the rest of the list is the right answer. Otherwise, the answer will be a list; the first element of s should be included, as should all of rest(s) except n; therefore the last line is

$$\underline{\text{build(first(s), REMBER(n, rest(s)))}}\ .$$

A useful little book is [Friedman 74], which talks about strategies for writing function definitions in this style.

EXERCISES 4.1

1. Write a function that takes two finite sets of integers, represented as above, and returns their intersection.

2. Same as Exercise 1, but return their union.

3. Write a function that takes two lists of the same length, $(l_1 \ldots l_n)$ and $(m_1 \ldots m_n)$, and returns the list $((l_1\ m_1) \ldots (l_n\ m_n))$.

4. Write a one-place function that returns true$^{\mathfrak{L}}$ if its argument is a list of atoms, and false$^{\mathfrak{L}}$ otherwise.

[4]Note that several different lists may represent the same set. (How?)
[5]See footnote 3.

5. Write a one-place function that takes a list and returns true$^{\mathcal{L}}$ if its argument is a legal representation of a finite set of integers, and false$^{\mathcal{L}}$ otherwise. You may assume that you have a one-place predicate number?$^{\mathcal{L}}$, which returns true$^{\mathcal{L}}$ iff its argument is an integer.

6. A matrix can be represented as a list as follows:

$$\begin{array}{c} ((a_{11} \ldots a_{1n}) \\ (a_{21} \ldots a_{2n}) \\ \vdots \\ (a_{n1} \ldots a_{nn})) \end{array}$$

Write a function that takes a matrix and transposes it.

4.2 REPRESENTING DATA TYPES AS LISTS

In the preceding section, we wrote functions called MEMBER and REMBER, which, we claimed, performed operations on finite sets of integers. But they do not work with sets; they work with lists. If we are to prove the correctness of these functions, we must make precise the connection between the data type

$$\langle \text{finite sets of integers, MEMBER, REMBER} \rangle$$

and the data structures (in this case, lists) that represent them.

One might think that the appropriate connection would be made by a function C ("coding") that takes a finite set of integers and gives its representation as a list. This will not work, since a finite set of integers may be represented by several different lists. The approach that seems to work is to answer two questions[6]:

(i) Which lists are legal representations of finite sets of integers?
(ii) Given a legal representation, what finite set of integers does it represent?

Letting L denote the set of lists (as before) and *Fin* denote the set of all finite sets of integers, we therefore specify the connection between L and *Fin* by

(i) a subset $Leg \subseteq L$ ("legal representations")
(ii) a function D: $Leg \rightarrow Fin$ ("decoding")

[6]This approach was introduced, in the context of arrays and assignments, in [Hoare 72] (see also [Wegbreit and Spitzen 76]). These papers will be more readable after we cover the material of Chapter 6. The set (i) is called the "invariant" of the representations in [Hoare 72]; the function (ii) is called the "abstraction function."

We do this in the usual fashion, by induction:

Leg
(i) $\mathrm{nil}^{\mathcal{L}} \in Leg$
(ii) if $l \in Leg$, $x \in \omega$, and $x \notin D(l)$, then $\mathrm{build}^{\mathcal{L}}(x,l) \in Leg$
(iii) nothing else

D
(i) $D(\mathrm{nil}^{\mathcal{L}}) = \varnothing$
(ii) if $l \in Leg$, $x \in \omega$, and $x \notin D(l)$, then $D(\mathrm{build}^{\mathcal{L}}(x,l)) = \{x\} \cup D(l)$
(iii) nothing else[7]

You should stare at this definition long enough to be convinced that it agrees with the informal definition given in Section 1. Let us state a useful lemma:

Lemma 4.2.1. *If $l \in Leg$ and $l \neq \mathrm{nil}^{\mathcal{L}}$, then $D(l) = \{\mathrm{first}^{\mathcal{L}}(l)\} \cup D(\mathrm{rest}^{\mathcal{L}}(l))$.*

PROOF. If $l \in Leg$ and $l \neq \mathrm{nil}^{\mathcal{L}}$, then l must have been put into Leg by rule (ii). □

We can now state and prove the correctness of REMBER:

Theorem 4.2.1. *If $I(\underline{\mathrm{n}}) \in \omega$ and $I(\underline{\mathrm{s}}) \in Leg$, then*

(i) $M(I, \underline{\mathrm{REMBER(n,s)}}) \in Leg$
(ii) $D(M(I, \underline{\mathrm{REMBER(n,s)}})) = D(I(\underline{\mathrm{s}})) - \{I(n)\}$

PROOF. The induction is on the construction of Leg in stages Leg_0, Leg_1, etc. The induction hypothesis is IH_k: If $I(\underline{\mathrm{s}}) \in Leg_k$ and $I(\underline{\mathrm{n}}) \in \omega$, then

(i) $M(I, \delta\underline{\mathrm{REMBER}}) \in Leg$ and
(ii) $D(M(I, \delta\underline{\mathrm{REMBER}})) = D(I(\underline{\mathrm{s}})) - \{I(\underline{\mathrm{n}})\}$.

Base Step: If $I(\underline{\mathrm{s}}) \in Leg_0$, then $I(\underline{\mathrm{s}}) = \mathrm{nil}^{\mathcal{L}}$ and

$$M(I, \delta\underline{\mathrm{REMBER}}) = M(I, \underline{\text{if eq?(s, nil) then nil else}} \ldots)$$
$$= M(I, \underline{\mathrm{nil}}) = \mathrm{nil}^{\mathcal{L}}$$

which satisfies (i) and (ii).

Induction Step: Assume IH_k and let $I(\underline{\mathrm{s}}) \in Leg_{k+1}$. If $I(\underline{\mathrm{s}}) \in Leg_k$, then we are done by IH_k. Otherwise $I(\underline{\mathrm{s}}) = \mathrm{build}^{\mathcal{L}}(x,l)$ where $l = \mathrm{rest}^{\mathcal{L}}(I(\underline{\mathrm{s}})) \in Leg_k$,

[7]Because of the clause $x \notin D(l)$ in line (ii) of the definition of Leg, this is a simultaneous inductive definition of Leg and D, as in Exercise 2.3.6, rather than a definition of Leg followed by a definition of D. But this should not be a cause for concern.

Note also the analogy between what we are doing now for data structures and what we did earlier for programs.

$x = \text{first}^{\mathcal{L}}(I(\underline{\text{s}})) \in \omega$, and $x \notin D(l)$. In particular, $I(\underline{\text{s}}) \neq \text{nil}^{\mathcal{L}}$. So, taking the first $\underline{\text{else}}$ clause, we get

$$M(I, \delta\underline{\text{REMBER}}) = M(I, \underline{\text{if eq?(first(s), n) then rest(s)}}$$
$$\underline{\text{else build(first(s), REMBER(n, rest(s)))}}\ )$$

There are two cases:

Case I:

$$I(\underline{\text{n}}) = x.$$

Then $M(I, \delta\underline{\text{REMBER}}) = \text{rest}^{\mathcal{L}}(I(\underline{\text{s}}))$. So part (i) of IH_{k+1} is established. Furthermore, since $x = \text{first}^{\mathcal{L}}(I(\underline{\text{s}})) \notin D(l)$,

$$D(l) = D(I(\underline{\text{s}})) - \{\text{first}^{\mathcal{L}}(I(\underline{\text{s}}))\} = D(I(\underline{\text{s}})) - \{I(\underline{\text{n}})\}$$

This establishes part (ii) of IH_{k+1} for this case.

Case II:

$$I(\underline{\text{n}}) \neq x.$$

Then

$$M(I, \delta\underline{\text{REMBER}})$$
$$= \text{build}^{\mathcal{L}}(\text{first}^{\mathcal{L}}(I(\underline{\text{s}})), M(I, \underline{\text{REMBER(n, rest(s))}}))$$

$$\boxed{\begin{array}{l} I'(\underline{\text{n}}) = I(\underline{\text{n}}) \\ I'(\underline{\text{s}}) = \text{rest}^{\mathcal{L}}(I(\underline{\text{s}})) = l \end{array}}$$

$$= \text{build}^{\mathcal{L}}(\text{x}, M(I', \delta\,\underline{\text{REMBER}}))$$

Since $l \in Leg_k$, IH_k applies, so we know that

(a) $M(I', \delta\underline{\text{REMBER}}) \in Leg$
(b) $D(M(I', \delta\underline{\text{REMBER}})) = D(l) - \{I(\underline{\text{n}})\}$

Now we must show that

(i) $\text{build}^{\mathcal{L}}(x, M(I', \delta\underline{\text{REMBER}})) \in Leg$,
(ii) $D(\text{build}^{\mathcal{L}}(x, M(I', \delta\underline{\text{REMBER}}))) = D(I(\underline{\text{s}})) - \{I(\underline{\text{n}})\}$

For the first part, we need to show $x \notin D(M(I', \delta\underline{\text{REMBER}}))$. But $D(M(I', \delta\underline{\text{REMBER}})) = D(l) - \{I(\underline{\text{n}})\}$ and we already knew $x \notin D(l)$. By this and (a), we conclude (i) using the definition of *Leg*.

Last,

$$D(\text{build}^{\mathcal{L}}(x, M(I', \delta\underline{\text{REMBER}})))$$

$$\begin{aligned} &= \{x\} \cup (D(l) - \{I(\underline{\text{n}})\}) && \text{(by Lemma 4.2.1 and IH}_k\text{)} \\ &= (\{x\} \cup D(l)) - \{I(\underline{\text{n}})\} && \text{(since } I(\underline{\text{n}}) \neq x\text{)} \\ &= D(I(\underline{\text{s}})) - \{I(\underline{\text{n}})\} && \text{(by definition of } D\text{).} \end{aligned}$$

□

Hortatory remark: This proof may have seemed complicated, but in reality we have little alternative: We must be sure that a function like REMBER returns a legal representation, so that other functions can use the value it returns, and we must of course be sure it gives a representation of the right answer. What we did was merely write down all the things we must have considered when we were writing the program.[8] Part of the complexity of the proof was due to the need to organize it; mechanical techniques exist for this [Morris and Wegbreit 77]. In any case, we have expended much effort on making the meanings of programs precise; should we let our precise programs work on imprecise and fuzzy data?

End of hortatory remark

Another reason for going through the precise specification of data type representations is that it allows representations of several data types to coexist in the data type of lists. For example, if $\mathbb{S}$ is some data type with universe S represented by $Leg_{\mathbb{S}}$ and $D_{\mathbb{S}}$, then we can talk about "finite sets of elements of S" represented as lists without repetition, as follows:

(i) $\text{nil}^{\mathcal{L}} \in Leg$ and $D(\text{nil}^{\mathcal{L}}) = \varnothing$
(ii) if $x \in Leg_{\mathbb{S}}$, $l \in Leg$, and $D_{\mathbb{S}}(x) \notin D(l)$, then $\text{build}^{\mathcal{L}}(x,l) \in Leg$ and $D(\text{build}^{\mathcal{L}}(x,l)) = \{D_{\mathbb{S}}(x)\} \cup D(l)$
(iii) nothing else

If data type $\mathbb{S}$, with universe S, is represented by $Leg_{\mathbb{S}}$ and $D_{\mathbb{S}}$, what is the representation of a function, say from some finite set of atoms to S? Such a function is a finite set of elements of $At \times S$, so we can use the finite set representation, as follows:

(i) $\text{nil}^{\mathcal{L}} \in Leg$ and $D(\text{nil}^{\mathcal{L}}) = \varnothing$
(ii) if $a \in At$, $x \in Leg_{\mathbb{S}}$, $l \in Leg$, and for no $(a', x') \in D(l)$ does $a' = a$, then
   (a) $\text{build}^{\mathcal{L}}((a\ x), l) \in Leg$, and
   (b) $D(\text{build}^{\mathcal{L}}((a\ x), l)) = \{(a, D_{\mathbb{S}}(x))\} \cup D(l)$
(iii) nothing else

Thus, if $a_1, \ldots, a_n$ are distinct atoms, and $x_1, \ldots, x_n \in Leg_{\mathbb{S}}$, then the list

$$((a_1\ x_1) \ldots (a_n\ x_n))$$

represents the function

$$\{(a_1, D_{\mathbb{S}}(x_1)), \ldots, (a_n, D_{\mathbb{S}}(x_n))\}$$

Such a list is called an *association list*, and this representation is called the *association list representation* of a function of finite domain.

[8] This is a useful time to reread the first epigraph at the front of the book.

We now need functions to build association lists and to apply them as functions. The following function takes two arguments: a list $\underline{\ell}_1 = (a_1 \ldots a_n)$ of distinct domain values, and a list $\underline{\ell}_2 = (x_1 \ldots x_n)$ of the corresponding values in the range, and returns the correct association list.

$$\underline{\delta\ \text{PAIRLIST}(\ell_1, \ell_2)}$$
$$= \underline{\text{if eq?}(\ell_1, \text{nil}) \text{ then nil}}$$
$$\underline{\text{else build } (\text{2LIST}(\text{first}(\ell_1), \text{first}(\ell_2)),}$$
$$\underline{\text{PAIRLIST}(\text{rest}(\ell_1), \text{rest}(\ell_2)))}$$

Theorem 4.2.2. *If $I(\underline{\ell}_1) = (a_1 \ldots a_n)$ and $I(\underline{\ell}_2) = (x_1 \ldots x_n)$ where $a_1, \ldots, a_n$ are distinct atoms and $x_1, \ldots, x_n \in Leg_{\mathbb{S}}$, then $M(I, \delta\underline{\text{PAIRLIST}}) \in Leg$ and $D(M(I, \delta\underline{\text{PAIRLIST}})) = \{(a_1, D_{\mathbb{S}}(x_1)), \ldots, (a_n, D_{\mathbb{S}}(x_n))\}$.*

PROOF. Left as an exercise. □

Given an association list $\underline{\ell}$, to apply it as a function to an atom a we need only to search it looking for an element $(a\ x)$; x is then (the representation of) the answer:

$$\underline{\delta\ \text{LOOKUP}(\text{a}, \ell)}$$
$$= \underline{\text{if eq?}(\ell, \text{nil}) \text{then error}}$$
$$\underline{\text{else if eq?}(\text{first}(\text{first}(\ell)), \text{a}) \text{ then 2ND}(\text{first}(\ell))}$$
$$\underline{\text{else LOOKUP}(\text{a}, \text{rest}(\ell))}$$

Theorem 4.2.3. *If $I(\underline{\ell}) \in Leg$ and $I(\underline{\text{a}}) \in domain(D(\underline{\ell}))$, then*

$$D(M(I, \delta\ \underline{\text{LOOKUP}})) = (D_{\mathbb{S}}(I(\underline{\ell})))(I(\underline{\text{a}}))$$

PROOF. Left as an exercise. Note that the right-hand side is the application of the function $D_{\mathbb{S}}(I(\underline{\ell}))$ to argument $I(\underline{\text{a}})$. □

Note that, in general, D must be a surjective function, so that every value in the data type has some representation.

Another important fact is that a data type may be represented in several different ways. Such alternative representations are explored in Exercises 4.2.5–4.2.6.

EXERCISES 4.2

1. State and prove the correctness of

(a) MEMBER
(b) UNION
(c) INTERSECTION

2. How would MEMBER need to be modified to deal with finite subsets of S for some other data type $\mathbb{S}$, as discussed in the text?

3. Prove Theorem 4.2.2.

4. Prove Theorem 4.2.3.

The following exercises deal with alternate representations of functions $F \to S$ where F is a finite subset of ω and S is represented by $(Leg_{\mathbb{S}}, D_{\mathbb{S}})$.

5. (a) Write down *Leg* and *D* for the representation of these functions as *sorted* association lists.
   (b) Write PAIRLIST and LOOKUP for this representation.
   (c) Prove the correctness of your versions of PAIRLIST and LOOKUP.

6. Such a function may also be represented as a *search tree* [Horowitz and Sahni 76, Section 9.1]. Repeat Exercise 5 for this representation.

7. Such a function may also be represented as $((n_1 \ldots n_k)(f(n_1) \ldots f(n_k)))$. Then δPAIRLIST(x_1, x_2) = 2LIST(x_1, x_2). Repeat Exercise 5 for this representation.

4.3 THE EVALUATOR FOR THE LANGUAGE OF EXPRESSIONS

In Section 3.7 we demonstrated how one could take an inductively defined function and write a program in the language of expressions to calculate it. In this section we shall do this for the most important inductively defined function we have seen so far—the meaning function M for a data type $\mathfrak{A}$. This is a more ambitious programming project than any we have seen so far, so we shall have to use some project management or "structured programming" techniques to keep things under control.

We shall have to represent a number of data types:

(i) the data type $\mathfrak{A}$ whose M we are trying to simulate
(ii) the data type of expressions
(iii) the data type of environments
(iv) the data type of functional environments

We shall use the data type $\mathfrak{L}$ of lists to represent all these data types. We can now write down the specifications for our project, as follows:

Specification: Given a data type $\mathfrak{A}$, create representations in $\mathfrak{L}$:

$(Leg_{\mathfrak{A}}, D_{\mathfrak{A}})$ of $\mathfrak{A}$

$(Leg_{\mathrm{Exp}}, D_{\mathrm{Exp}})$ of expressions over $\mathfrak{A}$

$(Leg_{\mathrm{ENV}}, D_{\mathrm{ENV}})$ of environments over $\mathfrak{A}$

$(Leg_{\mathrm{FENV}}, D_{\mathrm{FENV}})$ of functional environments over $\mathfrak{A}$

and create a functional environment δ in $\mathcal{L}$ such that whenever

$$I(\underline{\text{delta}}) \in Leg_{\text{FENV}}, \qquad I(\underline{\text{env}}) \in Leg_{\text{ENV}}, \qquad \text{and} \qquad I(\underline{\text{exp}}) \in Leg_{\text{Exp}},$$

then

$$M_{\mathcal{L}}(\delta, I, \underline{\text{MM(delta, env, exp)}}) \in \text{Leg}_{\mathcal{Q}}$$

and

$$D_{\mathcal{Q}}(M_{\mathcal{L}}(\delta, I, \underline{\text{MM(delta, exp, env)}})) = M_{\mathcal{Q}}(D_{\text{FENV}}(\underline{\text{delta}}), D_{\text{ENV}}(\underline{\text{env}}), D_{\text{Exp}}(\underline{\text{exp}}))$$

End of specification

Now, one might start out by writing down the representations for all the data types and then writing $\delta\underline{\text{MM}}$. Unfortunately, we do not know what the functions and predicates are for any of the data types, so this plan is not very practical. Instead, we shall write $\delta\underline{\text{MM}}$ first. As we write $\delta\underline{\text{MM}}$, we shall discover what functions and predicates we need. Only then can we specify our four data types and their simulations. This approach is called "top-down programming" or "stepwise refinement." We join E. W. Dijkstra in viewing the art of "programming (or problem solving) as the judicious postponement of decisions and commitments" [Dijkstra 72, p. 67].

In order to make our programs more mnemonic, we shall continue the use of mnemonic formal parameters and write

$$\delta\, \underline{\text{F(alpha, beta, gamma)}} = \underline{\text{if atom?(beta) then}} \ldots$$

instead of

$$r(\underline{\text{F}}) = 3; \qquad \delta\underline{\text{F}} = \underline{\text{if atom?}(\text{x}_2) \text{ then}} \ldots$$

We shall also supply $\mathcal{L}$ with a suitable supply of atoms. In particular, we assume that we have a distinct atom for each element of the language (see Section 3.2), i.e., each individual variable symbol, function symbol, predicate symbol, constant symbol, special symbol, and function variable symbol over $\mathcal{Q}$. We shall write down these atoms surrounded by quotation marks, e.g., "if", "x_1", etc. We shall also use a few other atoms (in quotes) as flags for the computations, e.g., "call", "ivs", etc.

At long last, we may now start writing code. *MM* is to take three arguments: the representation of a functional environment δ, the representation of an environment I, and the representation of an expression t. $\delta\underline{\text{MM}}$'s task is to figure out which of the five rules for M (from Table 3.5.1) ought to apply, and then send the data to a help function that will actually apply the rule. Each help function takes as its arguments just the data necessary to apply the rule.

```
δMM(delta, env, exp)
   =if eq?(RULE(exp),"const") then DO-CONST(exp)
     else if eq?(RULE(exp),"ivs") then DO-IVS(env,exp)
     else if eq?(RULE(exp),"appl") then DO-APPL(delta,env,exp)
     else if eq?(RULE(exp),"cond") then DO-COND(delta,env,exp)
     else if eq?(RULE(exp),"call")then DO-CALL(delta,env,exp)
     else error
```

Here RULE is a function that examines the representation of an expression and returns one of the five atoms shown, depending on which kind of expression it is. Notice that each of the five help functions DO-... receives exactly the information it needs according to Table 3.5.1: to evaluate a constant one needs neither delta nor env; to evaluate an ivs one does not need delta; the other rules need all three components.

We now look to see which functions we can write without making any decisions about data types. As it turns out, we have to make some decisions immediately:

DECISION 1. We will do call-by-value.

The call by value rule says: To evaluate $M(\delta, I, F(u_1, \ldots, u_n))$, evaluate each $M(\delta, I, u_i)$ and create a new environment I' in which each of the formal parameters of the fvs F is matched to the evaluated actual parameter; then evaluate the body of F in this new environment. Transcribing this description yields:

```
δDO-CALL(delta, env, exp)
   =MM(delta, NEWI(FORMAL-PARAMS(delta, FN-PART(exp)),
                   EVAL-LIST(delta, env, ARG-PART(exp))),
         BODY(delta, FN-PART(exp)))
```

FN-PART and ARG-PART ought to return the fvs and a list of the arguments in a function call expression; BODY and FORMAL-PARAMS are to find the body and the list of formal parameters for an fvs in a δ; NEWI is to build a new environment; and EVAL-LIST takes a list of expressions to be evaluated and returns a list of their values. EVAL-LIST is a routine recursion:

```
δ EVAL-LIST(delta, env, the-list)
   = if eq?(the-list, nil) then nil
       else build (MM(delta, env, first(the-list)),
                   EVAL-LIST(delta, env, rest(the-list)))
```

DECISION 2. We choose $\mathcal{A}$ to be $\mathcal{L}$, and we represent a list by itself.

This means that we need only deal with first, rest, build, atom?, and eq?. DO-APPL, and DO-COND are the functions that execute these functions.

DO-APPL looks at the function part of the expression and executes the appropriate function:

```
δDO-APPL(delta, env, exp)
    = if eq?(FN-PART(exp),"first") then
            first(MM(delta, env, first(ARG-PART(exp))))
        else if eq?(FNS7-PART(exp),"rest") then
                rest(MM(delta, env, first(ARG-PART(exp))))
        else build(MM(delta, env, first(ARG-PART(exp)))
                MM(delta, env, 2ND(ARG-Part(exp))))
```

Similarly, DO-COND looks at the predicate symbol of the expression, executes the appropriate test, and then evaluates the appropriate part of the expression:

```
δ DO-COND(delta, env, exp)
    = if eq?(IF-PART(exp),"atom?") then
            if atom?(MM(delta, env, PRED-ARG1(exp)))
              then MM(delta, env, THEN-PART(exp))
              else MM(delta, env, ELSE-PART(exp))
        else if eq?(MM(delta, env, PRED-ARG1(exp)),
                    MM(delta, env, PRED-ARG2(exp)))
                then MM(delta, env, THEN-PART(exp))
                else MM(delta, env, ELSE-PART(exp))
```

Here IF-PART extracts the predicate symbol, PRED-ARG1 and PRED-ARG2 extract the arguments to the predicate, and THEN-PART and ELSE-PART extract the then- and else- parts of the conditional.

We have now generated a healthy set of help functions to be written, which we list in Table 4.3.1. These functions all do manipulation on the representations of expressions, environments, etc. It is therefore time to decide on the representations. We will then write each function to do just what is expected of it by the functions that need it.

Since the largest number of functions manipulate expressions, we decide on their representation first.

DECISION 3 (Representation of Expressions). We shall represent each element of the language (see Section 3.2) by the corresponding quoted atom, e.g.,

the ivs x_1 is represented by the atom "x_1"

the function symbol f is represented by the atom "f"

etc.[9] Given an element of the language s, let $\lfloor s \rfloor$ denote the corresponding atom, e.g., $\lfloor x_1 \rfloor$ = "x_1", so $\lfloor s \rfloor$ is the representation of the element s.

We choose a representation for expressions in which each expression t has a unique representation t^*. It is therefore easy to define the representation function (—)* by an induction following the definition of an expression:[10]

t	t^*
c	("const" $\lfloor c \rfloor$)
v	("ivs" $\lfloor v \rfloor$)
$f(t_1, \ldots, t_n)$	("appl" $\lfloor f \rfloor$ $t_1^* \ldots t_n^*$)
if $p(u_1, \ldots, u_n)$ then t_1 else t_2	("cond" $\lfloor p \rfloor$ ($u_1^* \ldots u_n^*$) t_1^* t_2^*)
$F(t_1, \ldots, t_n)$	("call" $\lfloor F \rfloor$ $t_1^* \ldots t_n^*$)

Thus F(first(x_1),G(nil)) is represented by[11]

("call" "F"
    ("appl" "first" ("ivs" "x_1"))
    ("call" "G" ("const"( ))))

We may now write

δ RULE(exp) = first(exp)
δ DO-CONST(exp) = 2ND(exp)
δ FN-PART(exp) = 2ND(exp)
δ ARG-PART(exp) = rest(rest(exp))
δ IF-PART(exp) = 2ND(exp)
δ PRED-ARG1(exp) = first(3RD(exp))
δ PRED-ARG2(exp) = 2ND(3RD(exp))
δ THEN-PART(exp) = 4TH(exp)
δ ELSE-PART(exp) = 5TH(exp)

[9]This is not to be confused with constant symbol "x_1", which denotes the atom "x_1".

[10]Notice that (—)* is *postfix* notation for a function.

[11]While this representation may seem cumbersome, it is in fact typical of the intermediate structures built by a compiler. In such an intermediate form, all the parsing information is made explicit (see how this representation resembles Figure 3.4.1). Similarly, constant symbols may be evaluated, as was *nil* in this example.

TABLE 4.3.1 Help functions for the evaluator[a]

Function	Type of arguments	Type of result	Used by
RULE	exp	Atom	MM
DO-CONST	exp	List	MM
DO-IVS	env, exp	List	MM
NEWI	List-of-ivs, List-of-values,	Environment	DO-CALL
FORMAL-PARAMS	delta, fvs	List-of-ivs	DO-CALL
FN-PART	exp	fvs or function symbol	DO-CALL, DO-APPL
ARG-PART	exp	List-of-exp	DO-CALL, DO-APPL
BODY	delta, fvs	exp	DO-CALL
IF-PART	exp	Predicate symbol	DO-COND
PRED-ARG1	exp	exp	DO-COND
PRED-ARG2	exp	exp	DO-COND
THEN-PART	exp	exp	DO-COND
ELSE-PART	exp	exp	DO-COND

[a]A description of each help function is given in the text following the function that uses it.

DECISION 4 (Representing Environments). We represent an environment I as an association list, with individual variable symbols v represented by v^*. Thus the environment

$$\{(\underline{x_1}, \underset{\sim}{(FOO)}), (\underline{x_2}, \underset{\sim}{(\ )})\}$$

might be represented by the association list

$$\underset{\sim}{((("ivs"\ "x_1")\ (FOO))\ (("ivs"\ "x_2")(\ )))}$$

Now we can write

$$\delta\ \underline{\text{DO-IVS(env,exp)}} = \underline{\text{LOOKUP(exp,env)}}$$

and

$$\delta\ \underline{\text{NEWI(list-of-ivs, list-of-values)}} = \underline{\text{PAIRLIST(list-of-ivs, list-of-values)}}$$

where LOOKUP and PAIRLIST are the functions on association lists defined in Section 2.

DECISION 5 (Representation of the Functional Environment). In Table 3.5.1, a functional environment was a function δ: FVS$\rightarrow$Exp, associating an expression (the function body) with each function variable symbol (the function name). In our system, we also need to be able to retrieve the formal parameters, so a functional environment will be a function FVS$\rightarrow$ (IVS$^+ \times$Exp); thus δF is an ordered pair (formal-parameters, body). We represent such a function as an association list, representing each fvs F as the corresponding atom $\lfloor F \rfloor$ (which is what FN-PART retrieves). So if our functional environment contains

$$\delta\ \underline{\text{FOO(bar,a)}} = t_1$$

and

$$\delta\ \underline{\text{LOSE(x)}} = t_2$$

the functional environment might be represented by

$$\underline{((\text{"FOO"}\ ((\text{"bar"}\ \text{"a"})\ t_1^*))}$$
$$\underline{(\text{"LOSE"}\ ((\text{"x"})\ t_2^*))\ldots)}$$

where the ellipses elide any other function definitions to be represented.

Now we can write the last two help functions:

$$\delta\ \underline{\text{FORMAL-PARAMS(delta, fvs)}}$$
$$= \underline{\text{first(LOOKUP(fvs, delta))}}$$
$$\delta\ \underline{\text{BODY(delta, fvs)}} = \underline{\text{2ND(LOOKUP(fvs, delta))}}$$

This completes the construction of the evaluator MM.

EXERCISES 4.3

1. Modify the representation of environments to use v instead of v^*.

2. Modify DO-CALL to do call-by-name instead of call-by-value.

3. Translate the representation of expressions to the format used in Section 2.

4. Write an evaluator for the programming language of your choice.

*5. Prove that MM, as defined, meets the specifications given.

4.4 THE HALTING PROBLEM

In Section 3, we gave a demonstration that it was possible to write rather complex programs in our language of expressions. We wrote a function MM that took as input a representation δ^* of a functional environment δ, a representation I^* of our environment I, and a representation t^* of an expression t, and gave as its value the value of $M(\delta, I, t)$ if that value was defined. If, on the other hand, the evaluation of $M(\delta, I, t)$ by the rules of Table 3.5.1 failed to halt, then the function MM, which simulated those rules, would likewise fail to halt. One might propose writing a program which, given δ^*, I^*, and t^* as inputs, would tell whether or not $M(\delta, I, t)$

will halt. In this section we shall show that this halting problem is unsolvable: There is no such program, not only in our language of expressions but in any reasonable programming language.

Let us state the specifications on this program more precisely. The *halting problem* (for programs in some fixed programming language L), may be stated as follows:

> Given a program P (in language L) and input I, determine whether or not program P would halt if run on input I

A solution to the halting problem is an algorithm A (not necessarily written in L!) that takes as input P and I and always gives either true or false as its answer, depending on whether or not P halts on input I. Such an algorithm is called a *decision procedure* for the halting problem.

If L is any reasonable programming language, there is no program to solve the halting problem.

Before we prove this remarkable fact, it will be useful to retell a famous logical paradox, called the barber's paradox:

Hypothesis: There is a town with one barber who shaves exactly those people who do not shave themselves.

Question: Who shaves the barber?

Answer (after a fashion): The barber shaves himself if and only if he does not shave himself.

Conclusion: There is no such town.

Let us imagine now that we are to look at programs in some fixed programming language with some standard representation for their inputs.

Hypothesis: There exists a program A with two inputs, P and I, such that

(i) A halts on every input

(ii) $A(P,I)=\begin{cases} \text{TRUE} & \text{if program } P \text{ halts on input } I \\ \text{FALSE} & \text{if program } P \text{ does not halt on input } I \end{cases}$

If program A were available, a perverse programmer could write

Step 1. Read (x)
Step 2. If $A(x,x)=$TRUE then go to Step 2
Step 3. Halt

or something equivalent in the programming language under consideration. Let us call this program F. Notice that

$$F \text{ halts on input } x \text{ iff } x \text{ does not halt on input } x \qquad (*)$$

This property makes the program F reminiscent of the mythical barber, so we ask the obvious question:

Question: Does F halt on input F?

Substituting F for x in (*) gives:

Answer: F halts on F iff F does not halt on F.

Conclusion: There is no such program F.

But if there were a program A with the two properties given, then one could write program F. So

Theorem 4.4.1. *Program A does not exist.* □

This theorem applies to our language of expressions, but since it is not quite clear what is "program" and "input" for writing the function F, it is worthwhile to go through the details.

Theorem 4.4.2. *In the data type of lists, there is no functional environment δ_H such that for every environment I,*

$$M(\delta_H, I, \underline{\text{HALTS(delta, env, exp)}})$$

$$= \begin{cases} \underline{\text{true}} & \text{if } I(\underline{\text{delta}}) \text{ is the representation of a functional environment } \delta', I(\underline{\text{env}}) \text{ is the representation of an environment } I', \\ & \text{and } I(\underline{\text{exp}}) \text{ is the representation of an expression } t' \text{ such that } M(\delta', I', t') \text{ is well defined} \\ \underline{\text{false}} & \text{otherwise} \end{cases}$$

PROOF. Assume there is such a δ_H. Let $\underline{\text{F}}$ be an fvs not included in δ_H, and let δ_F be δ_H augmented by

$$\delta\,\underline{\text{F(x)}} = \underline{\text{if eq?(true, HALTS(x,}}$$
$$\underline{\text{1LIST(2LIST("x",x)),}}$$
$$\underline{\text{3LIST("call" "F" "x")))}}$$
$$\underline{\text{then F(x)}}$$
$$\underline{\text{else nil}}$$

The second argument to HALTS constructs the representation in $\mathcal{L}$ of the environment $\{(\underline{\text{x}}, l)\}$ where l is the value of F's argument. The third argument to HALTS is the representation in $\mathcal{L}$ of the expression $\underline{\text{F(x)}}$. Now let us evaluate $F(x)$ in an environment where $I(\underline{\text{x}})$ is the representation δ^* of some functional environment δ. F first asks whether

$M(\delta, \{(\underline{x}, \delta^*)\}, \underline{F(x)})$ halts. If it does, then F loops; if it does not, then F halts. In other words, for any functional environment δ,

$$M(\delta_F, \{(\underline{x}, \delta^*)\}, \underline{F(x)}) \text{ halts iff } M(\delta, \{(\underline{x}, \delta^*)\}, \underline{F(x)}) \text{ does not halt.}$$

Now choose δ_F for δ. We conclude that

$$M(\delta_F, \{(\underline{x}, \delta_F^*)\}, \underline{F(x)}) \text{ halts iff } M(\delta_F, \{(\underline{x}, \delta_F^*)\}, \underline{F(x)}) \text{ does not halt.}$$

Once we have δ_H, there is no escaping the paradox. So there must not be such a δ_H. □

This argument seems to depend on the fact that HALTS and F were written in the same language as the programs we were trying to check for halting. On the other hand, our programming experience, such as the writing of $\underline{\text{MM}}$, suggests that *any* effectively computable function (that is, a function that can be calculated by an algorithm) can be represented and computed in the language of expressions over lists. When the notion of "effectively computable" is made precise (which may be done in five or six standard ways, all of which come out exactly equivalent; see, e.g., [Brainerd and Landweber 74]) it indeed turns out that the class of functions computable by the language of expressions over lists is precisely the class of effectively computable functions.

EXERCISES 4.4

1. Show that the halting problem for conditionals (Section 3.4) *is* solvable.

2. Let $\mathfrak{A}$ be any data type such that A is finite. Show that the halting problem for $\mathfrak{A}$ is solvable by sketching a decision procedure for it.

3. Let us say a set x is *self-containing* if $x \in x$ and *nonself-containing* otherwise. Is the set of all nonself-containing sets self-containing or nonself-containing? Defend your answer.

5 The Language of Logic

5.1 LANGUAGES FOR FACTS AND QUESTIONS

At the end of Chapter 2, we proved the correctness of some algorithms. In each case, the proof used some *facts* about the data type. For example, the proof of the multiplication program used the facts:

$$x*0=0$$

$$x*(y+1)=x*y+x$$

Similarly, in Chapter 1, we asked *questions* about graphs. Given a directed graph $R \subseteq A \times A$ we could ask:

Is R a function?

Is R reflexive?

Is R transitive?

"R is transitive" is a possible *fact* about R.

As we write more complex algorithms, the correct functioning of those algorithms will generally depend on more complex facts about the data types being manipulated. Thus we need a language for the precise statement of facts and questions, just as we need a programming language for the precise statement of algorithms. This language is the language of *logic.* In this chapter, we shall develop the language of logic. Along the way, we shall see how the language of logic applies to the design of hardware. We shall also show how the language of logic may be used for the precise expression of arguments as well as facts.

5.2 THE LANGUAGE OF PROPOSITIONAL LOGIC

A complex question usually involves a large number of simpler questions. For example, if we are investigating a particular directed graph $R \subseteq A \times A$ the only "simple" questions we can ask are ones such as

Is $(a,b) \in R$?

Is $(a,c) \in R$?

Is $(c,b) \in R$?

and so on. Any other question, such as, "Is R transitive?" is answered by asking some of these simple questions and combining the answers in some way.

Propositional logic is the study of how simple questions, or *propositions*, are combined to form more complicated propositions. Given a complicated question, we want to know how we should combine the answers to the simpler propositions in order to answer the complicated question. Since every answer is either TRUE or FALSE, propositional logic boils down to the study of a data type whose values are just TRUE and FALSE. In fact, we shall largely forget until Section 6 the fact that we had intended to manipulate questions about a data type.

We define the data type $\mathcal{B}$ (of *Booleans*) as follows:

$$\mathcal{B} = \langle \{\mathrm{T},\mathrm{F}\}, \&^{\mathcal{B}}, \vee^{\mathcal{B}}, \supset^{\mathcal{B}}, \sim^{\mathcal{B}} \rangle$$

where

$$\begin{aligned} \&^{\mathcal{B}} &: \{\mathrm{T},\mathrm{F}\}^2 \to \{\mathrm{T},\mathrm{F}\} \\ \vee^{\mathcal{B}} &: \{\mathrm{T},\mathrm{F}\}^2 \to \{\mathrm{T},\mathrm{F}\} \\ \supset^{\mathcal{B}} &: \{\mathrm{T},\mathrm{F}\}^2 \to \{\mathrm{T},\mathrm{F}\} \\ \sim^{\mathcal{B}} &: \{\mathrm{T},\mathrm{F}\} \to \{\mathrm{T},\mathrm{F}\} \end{aligned}$$

are given by the following tables:

x	y	$\&^{\mathcal{B}}(x,y)$	$\vee^{\mathcal{B}}(x,y)$	$\supset^{\mathcal{B}}(x,y)$
F	F	F	F	T
F	T	F	T	T
T	F	F	T	F
T	T	T	T	T

x	$\sim^{\mathcal{B}}(x)$
F	T
T	F

Definition. A *well formed formula* (or *wff*) is a term (see Section 3.3) in the data type $\mathcal{B}$.

A wff represents a possible combination of elementary propositions. Therefore, the individual variable symbols are often called *propositional variables*. In keeping with conventional notation, we shall use $\underline{p}, \underline{q}, \underline{r}, \ldots$ as propositional variables instead of $\underline{x}, \underline{y}, \underline{z}, \ldots$. We also follow standard usage and use infix notation (see Section 1.1): we write

$$p \,\&\, q \qquad \text{instead of} \qquad \underline{\&(p,q)}$$

just as we usually write $2+3$ instead of $\underline{+(2,3)}$; we also dispense with underlining wherever possible. We use the standard technique of *precedence* to control order of evaluation in an infix term. Our precedence is

$\sim$	evaluate first
$\&$	
$\vee$	
$\supset$	evaluate last

Thus:

$$p \vee q \& r \qquad \text{means} \quad p \vee (q \& r)$$
$$\sim p \supset q \vee r \qquad \text{means} \quad (\sim p) \supset (q \vee r)$$

Repeated operators are evaluated left to right:

$$p \supset q \supset r \qquad \text{means} \quad (p \supset q) \supset r$$

The intention of the functions in $\mathcal{B}$ is as follows: $p \& q$ represents the complex proposition "p is true and q is true," and $\&^{\mathcal{B}}(x,y) = \mathrm{T}$ iff both x and y evaluate to T. $p \vee q$ represents the complex proposition "p is true or q is true," and $\vee^{\mathcal{B}}(x,y) = \mathrm{T}$ iff either x or y (or both) evaluates to T. $p \supset q$ represents the complex proposition "if p, then q" or "p implies q." If p is false, then $p \supset q$ is automatically true, since the condition p does not apply, but if p is true, then $p \supset q$ is true iff q is true.

Since a wff is just a term, we can use the techniques of Section 3.3 to evaluate wffs. An environment I is just a function $\mathrm{IVS} \to \{\mathrm{T}, \mathrm{F}\}$. We often think of an environment I as specifying a "possible world," e.g., "Consider a world where John loves Mary and Mary does not love John." A *truth table* is tabular format for evaluating a wff in all possible environments. Each line of a truth table corresponds to one combination of values of the variables of the wff. Each column is a wff which is a part of the wff being evaluated, starting with the propositional variables and ending with the entire wff. Tables 5.2.1 show some truth tables. Again, in keeping with the standard terminology in logic, we shall sometimes refer to an environment as an *interpretation*.

Since the value of $M(I,t)$ depends on $I(v)$ for the variables v in t, if the wff t evaluates to T on every line of the truth table, then it evaluates to T in every interpretation.

TABLES 5.2.1 Some truth tables

p	q	$p \vee q$	$\sim p$	$p \vee q \supset \sim p$
F	F	F	T	T
F	T	T	T	T
T	F	T	F	F
T	T	T	F	F

p	$\sim p$	$p \& \sim p$
F	T	F
T	F	F

p	q	r	$p \supset q$	$q \supset r$	$(p \supset q)\&(q \supset r)$	$p \supset r$	$(p \supset q)\&(q \supset r) \supset (p \supset r)$
F	F	F	T	T	T	T	T
F	F	T	T	T	T	T	T
F	T	F	T	F	F	T	T
F	T	T	T	T	T	T	T
T	F	F	F	T	F	F	T
T	F	T	F	T	F	T	T
T	T	F	T	F	F	F	T
T	T	T	T	T	T	T	T

By inspecting the truth tables, we therefore see that $(p \supset q)\&(q \supset r) \supset (p \supset r)$ is true in every interpretation. This is what one would expect; the wff says, "If p implies q and q implies r, then p implies r," which ought to be a true statement.

Definition. A wff t is a *tautology* iff for every I, $M(I,t)=\text{T}$. A wff t is a *contradiction* iff for every I, $M(I,t)=\text{F}$.

A tautology is true "in all possible worlds"; that is, it is true regardless of the truth or falsity of its propositional variables. Thus if p represents the statement, "the moon is made of green cheese," then the tautology $p \vee \sim p$ represents the statement:

either the moon is made of green cheese or the moon is not made of green cheese

which is true regardless of whether or not the moon is really made of green cheese. Similarly, a contradiction is false regardless of the truth or falsity of its propositional variables.

Definition. Two wffs t_1 and t_2 are *logically equivalent* (t_1 eq t_2) iff for every I, $M(I,t_1)=M(I,t_2)$.

Table 5.2.2 shows some pairs of logically equivalent wffs. Two logically equivalent wffs are just different ways of saying the same thing, that is, given any set of truth values for their elementary propositions, they both give the same answer.

TABLE 5.2.2 Some pairs of logically equivalent wffs

$p\&p$	eq	p
$p\vee p$	eq	p
$p\&(q\&r)$	eq	$(p\&q)\&r$
$p\vee(q\vee r)$	eq	$(p\vee q)\vee r$
$p\vee(q\&r)$	eq	$(p\vee q)\&(p\vee r)$
$p\&(q\vee r)$	eq	$(p\&q)\vee(p\&r)$
$\sim(p\vee q)$	eq	$\sim p\&\sim q$
$\sim(p\&q)$	eq	$\sim p\vee\sim q$
$\sim\sim p$	eq	p
$p\supset q$	eq	$\sim p\vee q$

Theorem 5.2.1. *If A and B are wffs, then A* eq *B iff $(A\supset B)\&(B\supset A)$ is a tautology.*

PROOF. Assume $A\,\text{eq}\,B$. In any environment I, either $M(I,A)=M(I,B)=\text{T}$ or $M(I,A)=M(I,B)=\text{F}$. If $M(I,A)=M(I,B)=\text{T}$, then

$$M(I,(A\supset B)\&(B\supset A))=\&^{\mathfrak{B}}(\supset^{\mathfrak{B}}(\text{T},\text{T}),\supset^{\mathfrak{B}}(\text{T},\text{T}))=\&^{\mathfrak{B}}(\text{T},\text{T})=\text{T}.$$

If $M(I,A)=M(I,B)=\text{F}$, then

$$M(I,(A\supset B)\&(B\supset A))=\&^{\mathfrak{B}}(\supset^{\mathfrak{B}}(\text{F},\text{F}),\supset^{\mathfrak{B}}(\text{F},\text{F}))=\&^{\mathfrak{B}}(\text{T},\text{T})=\text{T}.$$

So in every environment I, $M(I,(A\supset B)\&(B\supset A))=\text{T}$.

On the hand, assume A is not logically equivalent to B. Then there is some I such that $M(I,A)\neq M(I,B)$. If $M(I,A)=\text{T}$ and $M(I,B)=\text{F}$, then $M(I,A\supset B)=\text{F}$, and $M(I,(A\supset B)\&(B\supset A))=\text{F}$, so $(A\supset B)\&(B\supset A)$ is not a tautology. The case where $M(I,A)=\text{F}$ and $M(I,B)=\text{T}$ follows similarly. □

EXERCISES 5.2

1. Write the truth table for each of the following formulas and determine whether it is a tautology, a contradiction, or neither.

   (a) $(p\supset(p\supset p))\supset p$
   (b) $(p\supset\sim q)\supset(r\supset\sim q\vee p)$
   (c) $(p\vee r)\&(\sim p\vee q)\supset q\vee r$
   (d) $(p\supset q)\vee(q\supset p)$

2. Show

   (a) $\sim(p\wedge q)$ eq $\sim p\vee\sim q$
   (b) $p\supset q$ eq $\sim q\supset\sim p$

3. We say a wff A *logically implies* a wff B (in symbols, $A\models B$) iff for every I, if $M(I,A)=\text{T}$, then $M(I,B)=\text{T}$. For example, $(p\supset q)\&p\models q$. Show (by truth tables):

   (a) $p\supset q\models(q\supset r)\supset(p\supset r)$
   (b) $p\&(q\vee r)\models(p\&q)\vee r$

4. Consider the data type $\mathfrak{B}' = \langle\{T,F\}, \downarrow^{\mathfrak{B}'}\rangle$ where $\downarrow^{\mathfrak{B}'}$ is given by

x	y	$\downarrow^{\mathfrak{B}'}$
F	F	T
F	T	F
T	F	F
T	T	F

Give terms t_1 and t_2 in the data type $\mathfrak{B}'$ such that

(a) t_1 has the same truth table as $p \supset q$
(b) t_2 has the same truth table as $\sim p$

***5.** Describe an algorithm that given a wff t, produces a term t' in $\mathfrak{B}'$ that has the same truth table as t.

***6.** (De Morgan Duality) For any wff t, with no appearance of "$\supset$," define t' inductively as follows:

(i) if t is a variable, then $t' = \sim t$
(ii) if $t = \sim u$, then $t' = \sim u'$
(iii) if $t = \bigvee(u, v)$, then $t' = \&(u', v')$
(iv) if $t = \&(u, v)$, then $t' = \bigvee(u', v')$

Prove: For any t, $t \,\text{eq} \sim t'$.

7. Show that eq is reflexive, symmetric and transitive.

8. Show that $\models$ is reflexive and transitive but not symmetric.

5.3 SUBSTITUTION

In Table 5.2.2, we saw some particular pairs of wffs that were logically equivalent. Theorem 5.2.1, on the other hand, dealt with a *class* of pairs of wffs. $(A \supset B)\&(B \supset A)$ is not a wff; it stands for the class of all wffs of the *form* $(A \supset B)\&(B \supset A)$ for some wffs A and B. On the other hand, it greatly resembles a wff, and it would be useful to exploit that resemblance. We can make the resemblance precise by noting that a wff is of the form $(A \supset B)\&(B \supset A)$ iff it can be obtained from the wff $(p \supset q)\&(q \supset p)$ by substituting some wff A for p and some wff B for q. Our first task is to define substitution.

Definition. A *substitution* is a function σ from the set of propositional variables to the set of wffs. We define a function Subst, which takes as arguments a wff and a substitution and which gives the result of performing the substitution on the wff, as follows:

$$\begin{aligned}
&\mathrm{Subst}(v,\sigma) &&= \sigma(v) \quad \text{if} \quad v \in \mathrm{IVS}\\
&\mathrm{Subst}(\underline{\&(}\, t_1\underline{,} t_2\underline{)},\sigma) &&= \underline{\&(}\,\mathrm{Subst}(t_1,\sigma)\,\underline{,}\ \mathrm{Subst}(t_2,\sigma)\underline{)}\\
&\mathrm{Subst}(\underline{\vee(}\, t_1\underline{,} t_2\underline{)},\sigma) &&= \underline{\vee(}\,\mathrm{Subst}(t_1,\sigma)\,\underline{,}\ \mathrm{Subst}(t_2,\sigma)\underline{)}\\
&\mathrm{Subst}(\underline{\supset(}\, t_1\underline{,} t_2\underline{)},\sigma) &&= \underline{\supset(}\,\mathrm{Subst}(t_1,\sigma)\,\underline{,}\ \mathrm{Subst}(t_2,\sigma)\underline{)}\\
&\mathrm{Subst}(\underline{\sim(}\, t_1\underline{)},\sigma) &&= \underline{\sim(}\,\mathrm{Subst}(t_1,\sigma)\underline{)}
\end{aligned}$$

So a wff is of the form $(A\supset B)\&(B\supset A)$ iff it is $\mathrm{Subst}(\underline{(\mathrm{p}\supset\mathrm{q})\&(\mathrm{q}\supset\mathrm{p})},\sigma)$ for some σ.

Lemma 5.3.1 (Substitution lemma). *Let t be a wff, σ a substitution, and I an environment. Define an environment I' by $I'(v)=M(I,\sigma(v))$ for $v\in\mathrm{IVS}$. Then*

$$M(I',t)=M(I,\mathrm{Subst}(t,\sigma))$$

PROOF. By induction on the construction of t (cf. Ex. 2.3.8).

Base Step: If $v\in\mathrm{IVS}$, then

$$M(I',v)=I'(v)=M(I,\sigma(v))=M(I,\mathrm{Subst}(v,\sigma))$$

Induction Step: If $t=\bigcirc\underline{(}t_1\underline{,}\,t_2\underline{)}$ (where $\bigcirc$ stands for $\underline{\&}$, $\underline{\vee}$, or $\underline{\supset}$), then

$$\begin{aligned}
M(I',t)&=\bigcirc^{\mathfrak{B}}(M(I',t_1),M(I',t_2)) && \text{(defn. of } M)\\
&=\bigcirc^{\mathfrak{B}}(M(I,\mathrm{Subst}(t_1,\sigma)),M(I,\mathrm{Subst}(t_2,\sigma))) && \text{(IH)}\\
&=M(I,\bigcirc\underline{(}\mathrm{Subst}(t_1,\sigma)\,\underline{,}\ \mathrm{Subst}(t_2,\sigma)\underline{)}) && \text{(defn. of } M)\\
&=M(I,\mathrm{Subst}(\bigcirc\underline{(}t_1\underline{,}\,t_2\underline{)},\sigma)) && \text{(defn. of Subst)}\\
&=M(I,\mathrm{Subst}(t,\sigma))
\end{aligned}$$

The case for $t=\underline{\sim(}t_1\underline{)}$ follows similarly. □

This bit of sleight-of-hand pays off in the following theorem:

Theorem 5.3.1 (Tautology theorem). *Let t be a wff and σ be a substitution. Then,*

if t is a tautology, so is $\mathrm{Subst}(t,\sigma)$.

PROOF. Assume t is a tautology, and let I be any environment. We want to show $M(I,\mathrm{Subst}(t,\sigma))=\mathrm{T}$. Define a new environment I' by $I'(v)=M(I,\sigma(v))$. Then, by the substitution lemma, $M(I,\mathrm{Subst}(t,\sigma))=M(I',t)=\mathrm{T}$, since t is a tautology. □

Thus we may assert that any wff of the form $A\vee\sim A$ is a tautology, no matter how many variables appear in A, without constructing a complete truth table: We observe that $p\vee\sim p$ is a tautology and apply the tautology theorem. Similarly, the following theorem says that a single logical equivalence "lifts" to an assertion about a class of wffs:

Theorem 5.3.2 (Substitution theorem). *If* t_1 eq t_2, *then*

$$\text{Subst}(t_1,\sigma) \text{ eq } \text{Subst}(t_2,\sigma).$$

PROOF. Assume t_1 eq t_2 and let I be any environment. Let I' be given by $I'(v)=M(I,\sigma(v))$. Then

$$M(I,\text{Subst}(t_1,\sigma))=M(I',t_1)=M(I',t_2)=M(I,\text{Subst}(t_2,\sigma)). \qquad \square$$

So not only is $p\vee(q\vee r)$ eq $(p\vee q)\vee r$, but also for any wffs A, B, and C, $A\vee(B\vee C)$ eq $(A\vee B)\vee C$.

The following theorem says that you can replace any subwffs appearing in a wff by logically equivalent subwffs and get a wff that is logically equivalent to the original.

Theorem 5.3.3 (Replacement theorem). *Let* σ_1 *and* σ_2 *be substitutions such that for any* $v\in$IVS, $\sigma_1(v)$ eq $\sigma_2(v)$. *Then for any wff* t,

$$\text{Subst}(t,\sigma_1) \text{ eq } \text{Subst}(t,\sigma_2)$$

PROOF. Let I be any environment. Let I' be given by $I'(v)=M(I,\sigma_1(v))$. Since $\sigma_1(v)$ eq $\sigma_2(v)$, it is a fact that for each $v\in$IVS, $I'(v)=M(I,\sigma_2(v))$. So, using the substitution lemma twice,

$$M(I,\text{Subst}(t,\sigma_1))=M(I',t)=M(I,\text{Subst}(t,\sigma_2)). \qquad \square$$

For example, $p\supset q$ eq $\sim q\supset\sim p$, so $(q\supset p)\&(p\supset q)$ eq $(q\supset p)\&(\sim q\supset\sim p)$, using the replacement theorem with $t=(q\supset p)\&r$, and σ_1 and σ_2 given by

v	$\sigma_1(v)$	$\sigma_2(v)$
p	p	p
q	q	q
r	$p\supset q$	$\sim q\supset\sim p$

EXERCISES 5.3

Show that for any wffs P, Q, R, S, and U, the following are tautologies:

1. $(P\supset Q)\&(\sim P\supset Q)\supset Q$

2. $P\&(Q\vee R)\supset(P\&Q)\vee(P\&R)$

3. $(P\&Q)\vee(P\&R)\supset P\&(Q\vee R)$

4. $(P\supset Q)\supset(\sim Q\supset\sim P)$

Show:

5. $P\&Q \supset R$ eq $(\sim P \vee \sim Q) \vee R$

6. $P \vee \sim Q \vee (R \supset S) \supset U$ eq $P \vee \sim Q \vee (\sim S \supset \sim R) \supset U$

7. $P \vee (Q\&R)$ eq $(P \vee Q)\&(P \vee R)$

5.4 DISJUNCTIVE NORMAL FORM

If we can use many different, logically equivalent wffs to ask the same question, then it is reasonable to wonder whether we really needed all those different wffs in the first place. Is there, perhaps, some set of "moderately simple" wffs that will suffice to express any possible combination of propositional variables? The answer is, "Yes." In this section we shall define such a class, the class of wffs in *disjunctive normal form*, and show that every wff is logically equivalent to a wff in disjunctive normal form. We shall first need some auxiliary definitions.

Definition. A *literal* is a propositional variable or its negation.

Definition. An *elementary conjunct* is defined as follows:

(i) every literal is an elementary conjunct
(ii) if t_1 and t_2 are elementary conjuncts, then so is $\&(t_1, t_2)$
(iii) nothing else

A typical elementary conjunct in infix notation is $p\&\sim q\&r\&s$. Because & is associative, it makes no difference how the literals are grouped. Similarly, it makes no difference in what order the literals are written down, or whether any are repeated.

Definition. The set of wffs in *disjunctive normal form* (DNF) is defined as follows:

(i) every elementary conjunct is in disjunctive normal form
(ii) if t_1 and t_2 are in disjunctive normal form, then $\vee(t_1, t_2)$ is in disjunctive normal form
(iii) nothing else

A typical wff in DNF is $(p\&\sim q\&r) \vee (p\&q\&s) \vee (\sim p\&s)$. Notice that not every propositional variable need appear in every elementary conjunct. A wff in DNF is the "or" of a set of elementary conjuncts, each of which is the "and" of a set of literals.

Theorem 5.4.1 (Disjunctive normal form theorem). *To every wff A there is a logically equivalent wff B in disjunctive normal form.*

PROOF. Let A be any wff. Let U be the finite set of propositional variables appearing in A. For each environment I: $U \to \{T, F\}$, collect the following set of literals: if $M(I,v)=T$, take v; if $M(I,v)=F$, take $\sim v$. Form an elementary conjunct t_I by "and"-ing together the literals collected for environment I. Notice that $M(I',t_I)=T$ iff $I'=I$, because if $I'(v)\neq I(v)$, then the corresponding literal in I' will turn false. Now form a DNF wff B by or-ing all of the t_I such that $M(I,A)=T$. B evaluates to true iff one of the elementary conjuncts evaluates to true, which happens iff the environment is one of the ones that made A true. So A eq B. This works fine unless A is a contradiction, in which case there are no elementary conjuncts to put in B (so B is not a wff). In that case, set B to be $p\&\sim p$, since all contradictions are logically equivalent. □

EXAMPLE 1. Let $A=((p\&q)\vee(\sim r\&\sim q))\&(p\vee q\vee r)$. The truth table for A is

p	q	r	$p\&q$	$\sim r\&\sim q$	$(p\&q)\vee(\sim r\&\sim q)$	$p\vee q\vee r$	A	
F	F	F	F	T	T	F	F	
F	F	T	F	F	F	T	F	
F	T	F	F	F	F	T	F	
F	T	T	F	F	F	T	F	
T	F	F	F	T	T	T	T	(1)
T	F	T	F	F	F	T	F	
T	T	F	T	F	T	T	T	(2)
T	T	T	T	F	T	T	T	(3)

Looking at lines (1), (2), and (3), we get

$$B=(p\&\sim q\&\sim r)\vee(p\&q\&\sim r)\vee(p\&q\&r)$$

There may be more than one wff in DNF that is logically equivalent to the given wff. Another DNF for the A of Example 1 would be $(p\&\sim q\&\sim r)\vee(p\&q)$.

EXAMPLE 2. Let $A=(p\vee q\vee r)\&(\sim p\vee\sim q)$.

p	q	r	$p\vee q\vee r$	$\sim p\vee\sim q$	$(p\vee q\vee r)\&(\sim p\vee\sim q)$	
F	F	F	F	T	F	
F	F	T	T	T	T	(1)
F	T	F	T	T	T	(2)
F	T	T	T	T	T	(3)
T	F	F	T	T	T	(4)
T	F	T	T	T	T	(5)
T	T	F	T	F	F	
T	T	T	T	F	F	

Since there are five lines where A is true, there will be five elementary conjuncts in the DNF constructed according to the proof. The wff is

$$\begin{aligned}&(\sim p\&\sim q\&r)\\ \vee&(\sim p\&q\&\sim r)\\ \vee&(\sim p\&q\&r)\\ \vee&(p\&\sim q\&\sim r)\\ \vee&(p\&\sim q\&r)\end{aligned}$$

EXERCISES 5.4

1. Convert each wff in Exercise 5.2.1 to disjunctive normal form.

2. (Conjunctive normal form): Define a *clause* to be $A_1 \vee \cdots \vee A_n$ where the A_i's are literals. Use Exercise 5.2.6 and the disjunctive normal form theorem to prove: Every wff is logically equivalent to a wff of the form $B_1 \& \cdots \& B_m$, where the B_i's are clauses.

5.5 APPLICATIONS OF PROPOSITIONAL LOGIC

A major application of propositional logic in computer science lies in the area of hardware design. Since almost all computers represent their data internally as strings of bits, each bit having one of two values (0 and 1), and since propositional logic is the study of a data type $\mathfrak{B}$ with two possible values (F and T) one might expect that propositional logic might be useful for manipulating bits. It turns out that it is simple to build electronic circuits to compute $\&^{\mathfrak{B}}$, $\vee^{\mathfrak{B}}$, and $\sim^{\mathfrak{B}}$. This allows us to "compile" wffs into hardware: that is, to translate a wff into an electronic circuit that will evaluate it.

For the remainder of this section we shall use the bit 0 to represent F and the bit 1 to represent T. The circuits used to compute $\&^{\mathfrak{B}}$, $\vee^{\mathfrak{B}}$, and $\sim^{\mathfrak{B}}$ are called *gates*. Figure 5.5.1 shows the symbols used to represent and-gates, or-gates, and not-gates. Since $\&^{\mathfrak{B}}$ and $\vee^{\mathfrak{B}}$ are associative and commutative, one may have an and-gate with any number of inputs (limited only by what the manufacturers have decided to produce!); not-gates, on the other hand, are restricted to one input.

Now it is a simple matter to draw a circuit to evaluate any wff. Figure 5.5.2 shows two examples.

The disjunctive normal form theorem says that every wff can be realized by a circuit consisting of three layers: first a layer of not-gates; next a layer of and-gates, one for each elementary conjunct; and last a single or-gate. This is shown in Figure 5.5.3.

Generating a circuit diagram from a given wff is fairly trivial. It is more interesting to try to select the "best" of the wffs logically equivalent to the given wff, e.g., the one that requires fewest gates. This process, called

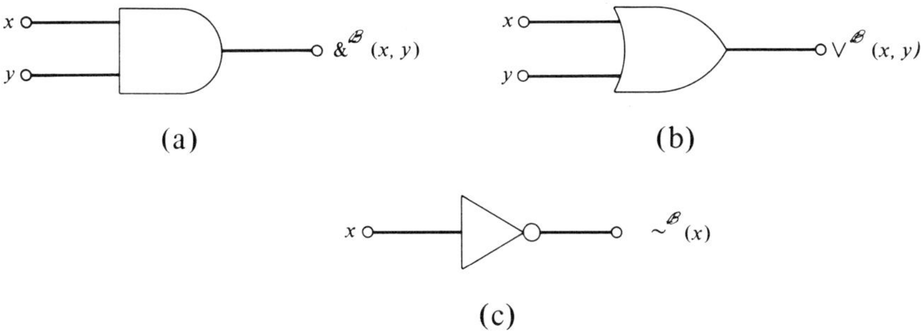

FIGURE 5.5.1 Symbols for gates. (a) and-gate, (b) or-gate, (c) not-gate.

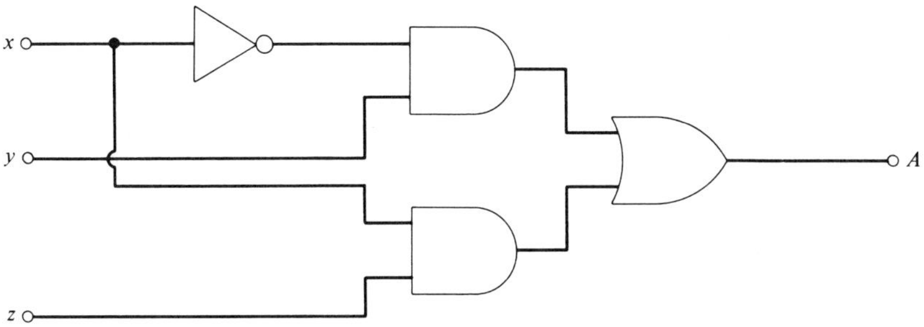

$A = (\sim x \,\&\, y) \vee (x \,\&\, z)$

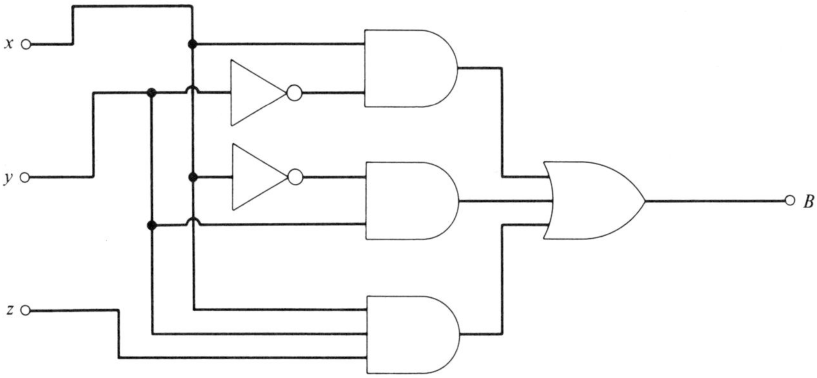

$B = (x \,\&\, \sim y) \vee (\sim x \,\&\, y) \vee (x \,\&\, y \,\&\, z)$

FIGURE 5.5.2

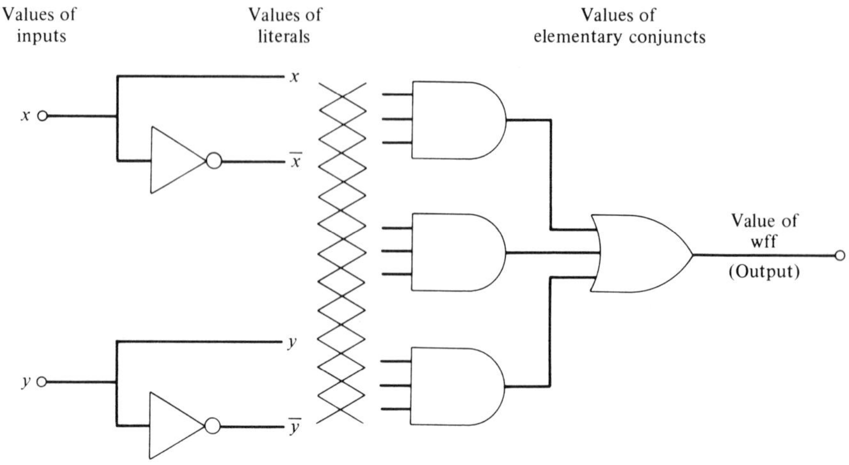

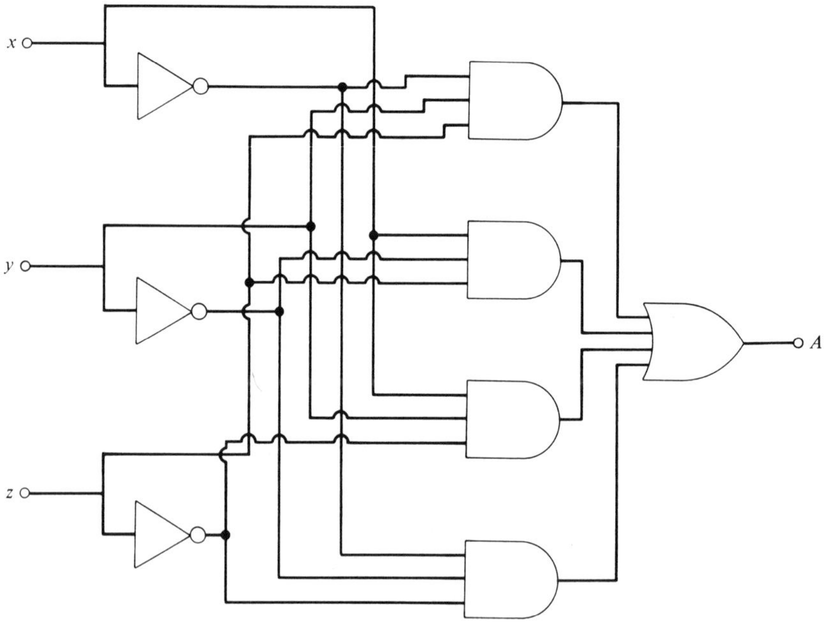

$A = (\sim x \;\&\; y \;\&\; z) \vee (x \;\&\; \sim y \;\&\; z) \vee (x \;\&\; y \;\&\; \sim z) \vee (\sim x \;\&\; \sim y \;\&\; \sim z)$

FIGURE 5.5.3 Circuits for disjunctive normal form.

boolean minimization, is the subject of an extensive literature (see [Korfhage 74, Section 8.5] or [Prather 76, Sections 4.8–4.9]).

Another more interesting logical circuit, shown in Figure 5.5.4, is called a *flipflop*. When $S=R=0$, this circuit may have $X=0$ and $Y=1$ or $X=1$ and $Y=0$—either set of values is consistent. If $S=1$ and $R=0$, then $X=0$ and $Y=1$; if $R=1$ and $S=0$, then the only consistent set of values is $X=1$ and $Y=0$. Let us imagine, then, letting $R=1$ for a brief period of time, and then letting $R=S=0$. The flipflop will go into the state given by $Y=0$. When R returns to 0, the flipflop will stay in the state $Y=0$ indefinitely, until a 1 is applied to input S. Then it will go into the state $Y=1$ and stay there until another 1 appears at R. This sequence of events is shown in Figure 5.5.5. Thus the flipflop "remembers" which of its input lines, R or S, was the last to receive a 1 (compare times t_1, t_3, and t_5 in Figure 5.5.5). The circuit diagram symbol for a flipflop is shown in Figure 5.5.6. A 1 on the R, or *reset*, input causes Q to become 0; a 1 on the S, or *set*, input causes Q to become 1. In Figure 5.5.6, Q corresponds to output Y in Figure 5.5.4; output $\overline{Q}$ corresponds to output X in Figure 5.5.4.

FIGURE 5.5.4 A flipflop.

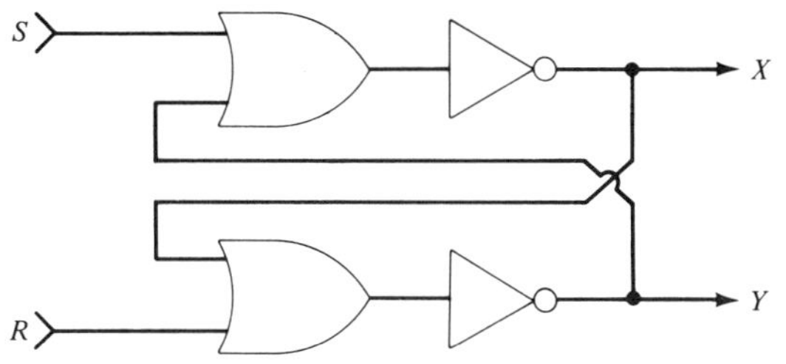

FIGURE 5.5.5 A sequence of operations on a flipflop.

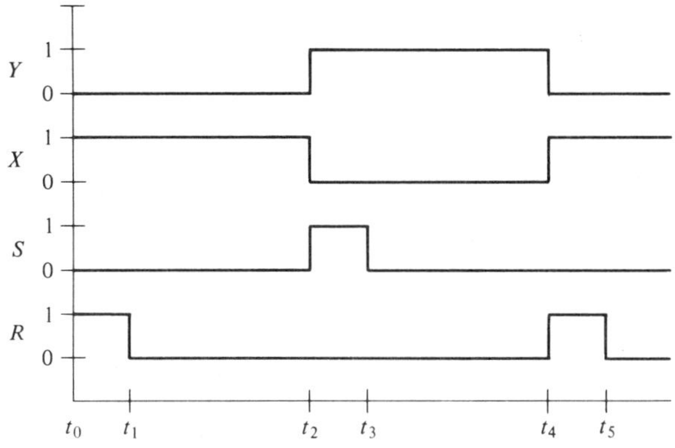

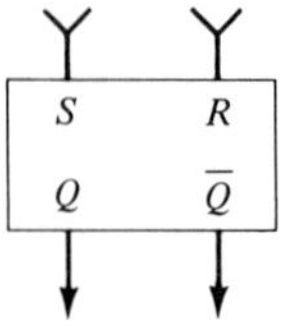

FIGURE 5.5.6 Symbol for a flipflop.

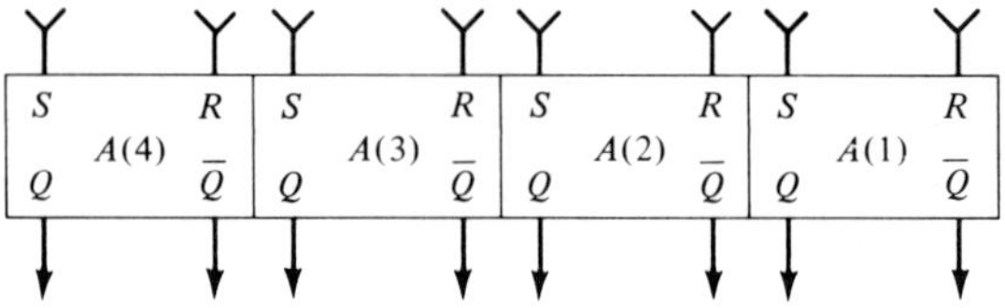

FIGURE 5.5.7 A four-bit register.

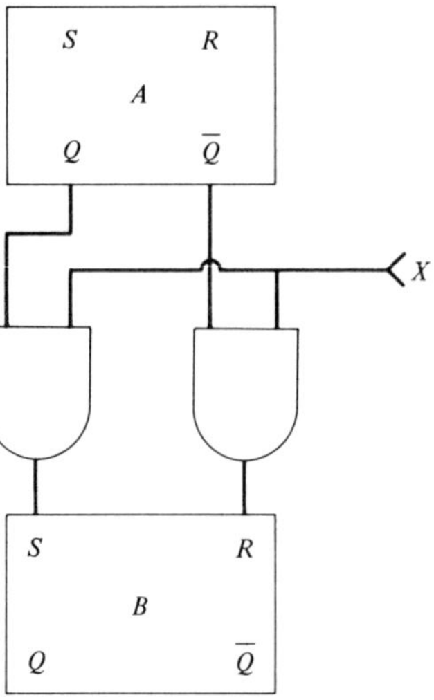

FIGURE 5.5.8 A transfer: $B \leftarrow A \,\&\, X$.

We say that the flipflop has *memory*, with a capacity of one bit: It remembers whether the last 1 appeared on input S or input R.[1] A computer register, such as an accumulator or program counter, is usually implemented as an array of flipflops. Figure 5.5.7 shows a four-bit register.

[1]This phenomenon arises because the flipflop circuit has something our previous circuits did not—a feedback loop. Just as in the case of programs, the addition of loops marks a major leap in complexity and expressive power. There are several flavors of flipflop that are actually used in circuits; the one we have described, called an *RS-flipflop*, is the simplest.

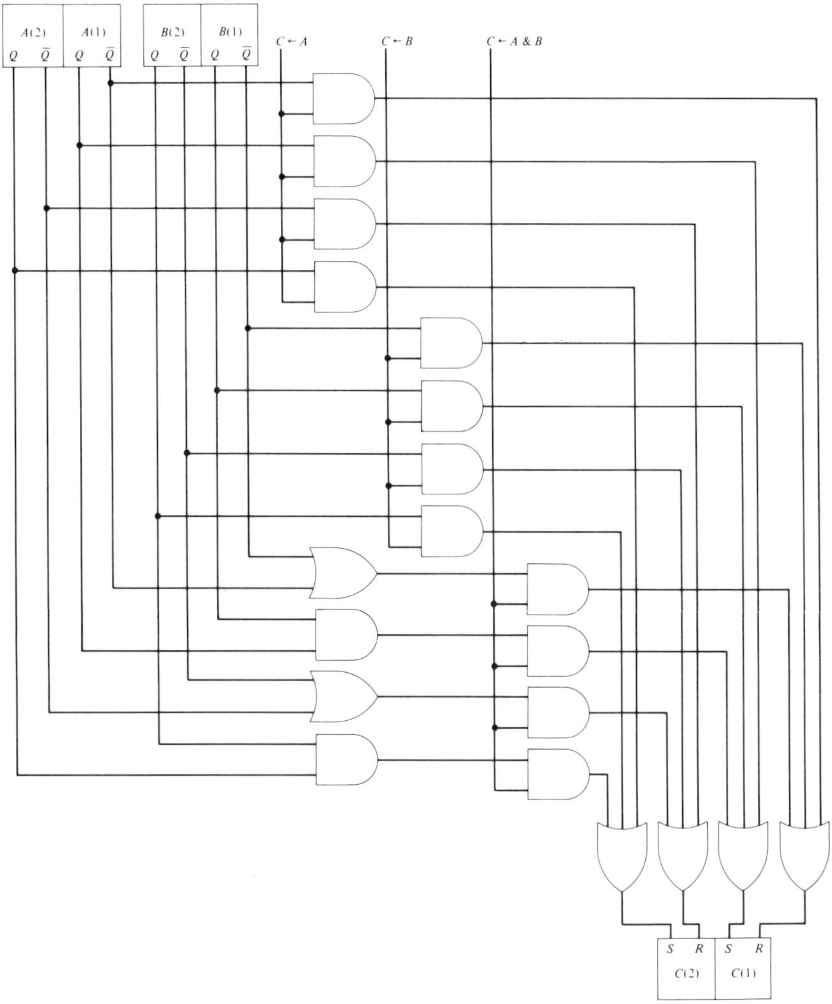

FIGURE 5.5.9 A logical unit with three operations.

Networks of gates are used to transfer information between flipflops. If a single 1 bit is sent to input X of the circuit in Figure 5.5.8, then flipflop B will be set to the contents of flipflop A. The circuit in Figure 5.5.9 performs three different functions: $C \leftarrow A$, $C \leftarrow B$, and $C \leftarrow A \& B$ on three two-bit registers.[2] The logical portion of a central processing unit consists of a large number of similarly simple circuits.

[2]Why does one of the groups of gates in Figure 5.5.9 have two and-gates and two or-gates?

EXERCISES 5.5

1. Design a circuit to compute each of the wffs in Exercise 5.2.1.

2. A *half adder* is a circuit that takes two one-bit numbers and computes their sum and carry:

x	y	s	c
0	0	0	0
0	1	1	0
1	0	1	0
1	1	0	1

Design a circuit for a half adder.

3. To cut down on the proliferation of *not*-gates, the following abbreviation is used:

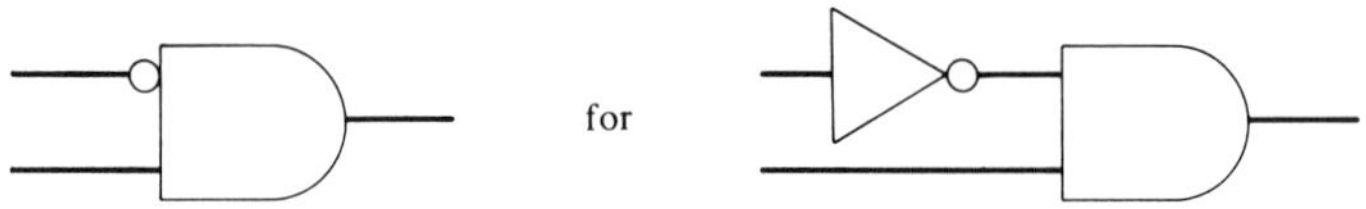

and similarly for any combination of negated and nonnegated inputs, and for or-gates as well. Draw a half adder circuit using this abbreviation.

4. A *full adder* is a circuit that takes three bits and outputs the binary sum and carry. Draw a circuit for a full adder. You may use the half adder as a building block.

5. Using full and half adders, design a circuit to add two two-bit numbers.

5.6 THE LANGUAGE OF FIRST-ORDER LOGIC

We now return to our original goal, namely, the development of a language to express facts ab‚ ut a data type. Our plan is as follows. First, we shall present a language for the "elementary propositions" in a data type. Next, we shall show how to combine these elementary propositions to form more complex propositions. Not only shall we use the boolean combinations studied so far, but also other ways of combining propositions will turn out to be useful and natural.

Assume that we are working in a data type

$$\mathcal{Q} = \langle A, f_1^{\mathcal{Q}}, \ldots, f_n^{\mathcal{Q}}, p_1^{\mathcal{Q}}, \ldots, p_m^{\mathcal{Q}}, c_1^{\mathcal{Q}}, \ldots, c_r^{\mathcal{Q}} \rangle.$$

Our goal is to define the language of *first-order formulas* over $\mathcal{Q}$. We shall do this in a now-familiar manner: We shall define inductively the set of formulas together with an inductive definition of a meaning function M. In this case the meaning function will take as arguments an environment[3] and a formula and will return either T or F.[4] In keeping with conventional notation, we shall refer to environments as *interpretations*. We shall again utilize the language of terms over $\mathscr{Q}$, defined in Section 2.3.

Definition. The set of *first-order formulas* over $\mathscr{Q}$ is the set of strings defined as follows:

(1) if p is an n-place predicate symbol, and $t_1,\ldots,t_n$ are terms, then $p\underline{(}t_1\underline{,}\ldots\underline{,}t_n\underline{)}$ is a formula

(2.1) if G and H are formulas, then $\underline{\&}GH$ is a formula

(2.2) if G and H are formulas, then $\underline{\vee}GH$ is a formula

(2.3) if G and H are formulas, then $\underline{\supset}GH$ is a formula

(2.4) if G is a formula, then $\underline{\sim}G$ is a formula

(3.1) if G is a formula, and v is an individual variable symbol, then $\underline{\exists}vG$ is a formula

(3.2) if G is a formula and v is an individual variable symbol, then $\underline{\forall}vG$ is a formula

(4) nothing else

As was true for the language of expressions, it will be convenient to have names for the formulas that are introduced by each rule. Table 5.6.1 gives these names.

Again, we shall often write our formulas in infix notation (except for quantifiers), using the precedence

$\forall$, $\exists$	evaluate first
$\sim$	
&	
$\vee$	
$\supset$	evaluate last

Thus, $\underline{(\forall x)(\text{is1?}(x) \supset (\exists y)[x = \text{push1}(y)])}$ is a typical formula.

Before defining the function M, let us see what each of these kinds of formulas *ought* to mean. We can then write down the appropriate rule for M. What is M supposed to do? It gets a formula G and an interpretation I and it is supposed to reply whether formula G is true or false in interpretation I.

Let us first consider atomic formulas. An atomic formula looks like $p\underline{(}t_1\underline{,}\ldots\underline{,}t_n\underline{)}$. Now each term t_i refers to a particular element of the set A of values in the data type, namely, $M(I,t_i)$ (where M is the term evaluator from Section 3.3). The predicate symbol p of course refers to the predicate

[3]Recall that an environment or interpretation is just a function I: IVS$\rightarrow A$.
[4]Or TRUE or FALSE, if you prefer.

TABLE 5.6.1

A formula introduced by rule	Is called a (an)
(1)	Atomic formula
(2.1)	Conjunction
(2.2)	Disjunction
(2.3)	Implication
(2.4)	Negation
(3.1)	Existential
(3.2)	Universal

$p^{\mathcal{Q}}$: $A^n \rightarrow \{T, F\}$. So it is reasonable to say that $p\underline{(}t_1\underline{,}\ldots\underline{,}t_n\underline{)}$ asks the question, "Is $p^{\mathcal{Q}}$ true at $M(I,t_1),\ldots,M(I,t_n)$?" Hence we write

$$M(I, p\,\underline{(}\,t_1\,\underline{,}\,\cdots\,\underline{,}\,t_n\,\underline{)}\,) = p^{\mathcal{Q}}(M(I,t_1),\ldots,M(I,t_n))$$

An atomic formula asks a question about the data type and is not a combination of simpler questions. Atomic formulas are thus the "elementary propositions" about data types.

Rules (2.1) through (2.4) are easy; they are just the boolean combinations.

Rules (3.1) and (3.2) are little more subtle. These operations are called *quantifiers*. Let us write down the entire definition of M and then attempt to explain them. As usual, we need an auxiliary definition:

Definition. If I and I' are interpretations, and v is an individual variable symbol, then I and I' are *equivalent modulo* v (we write $I \underset{v}{\equiv} I'$) iff for every individual variable symbol w, if $w \neq v$, then $I(w) = I'(w)$.

In other words, I and I' are equivalent modulo v if they agree everywhere except perhaps at v. For example, let us assume we are working in the real numbers and that there are only two variables to worry about, called $\underline{\mathrm{x}}$ and $\underline{\mathrm{y}}$. Let $I(\underline{\mathrm{x}}) = 2$ and $I(\underline{\mathrm{y}}) = -1$. Then $\{I' | I' \underset{\mathrm{y}}{\equiv} I\}$ is shown in Figure 5.6.1. Now we may write down the definition of M in Table 5.6.2.

TABLE 5.6.2 Meaning function for first-order formulas

(1)	$M(I, p\,\underline{(}\,t_1\,\underline{,}\,\cdots\,\underline{,}\,t_n\,\underline{)}\,) = p^{\mathcal{Q}}(M(I,t_1),\ldots,M(I,t_n))$
(2.1)	$M(I, \underline{\&}GH) = \&^{\mathcal{B}}(M(I,G), M(I,H))$
(2.2)	$M(I, \underline{\vee}GH) = \vee^{\mathcal{B}}(M(I,G), M(I,H))$
(2.3)	$M(I, \underline{\supset}GH) = \supset^{\mathcal{B}}(M(I,G), M(I,H))$
(2.4)	$M(I, \underline{\sim}G) = \sim^{\mathcal{B}}(M(I,G))$
(3.1)	$M(I, \underline{\exists}vG) = \mathrm{T}$ iff for some $I', I' \underset{v}{\equiv} I$ and $M(I',G) = \mathrm{T}$
(3.2)	$M(I, \underline{\forall}vG) = \mathrm{T}$ iff for every I' such that $I' \underset{v}{\equiv} I, M(I',T) = \mathrm{T}$

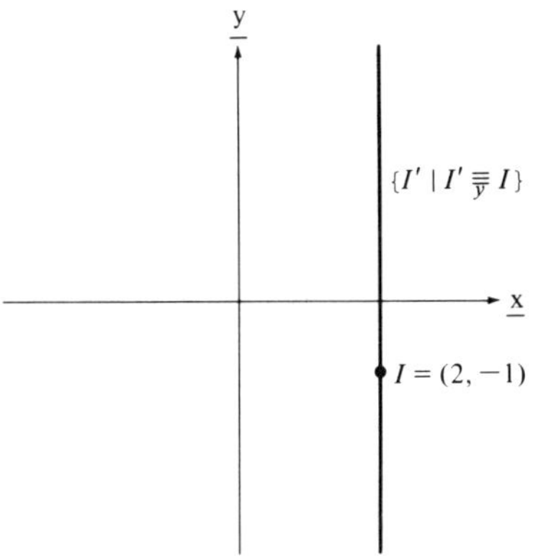

FIGURE 5.6.1

In Rule (3.1), $\underline{\exists} vG$ is read "*there exists* a v such that G." In other words, $\underline{\exists} vG$ should be true in interpretation I iff we can choose a value for v that makes G true. Since we have no stipulations about the other variables in G, we must leave them fixed. We can capture the requirements of both these sentences by saying that $\underline{\exists} vG$ should be true iff there exists some I', differing from I at v only, such that $M(I',G)=\mathrm{T}$. This is precisely what Rule (3.1) says.

In Rule (3.2), $\underline{\forall} vG$ is read "*for all* v, G." That is, $\underline{\forall} vG$ is to be true at I iff no matter how we vary the value of v, G will stay true. Again, all the other variables stay fixed, so we are again dealing with interpretations I' that are the same as I except possibly at v. In these terms, we say that $\underline{\forall} vG$ is true at I iff G is true at *all* of these I'. Again, this is just Rule (3.2) in Table 5.6.2.

Some more insight into the quantifiers may be obtained by looking at Figure 5.6.1. Again let $I=\{(\underline{x},2),(\underline{y},-1)\}$, and let G be any formula. Then $\underline{\exists} yG$ is true at I iff G is true at *some* point I' on the vertical line. $\underline{\forall} yG$ is true at I iff G is true at *every* point I' on the vertical line. So, just as the boolean connectives provide a way to combine the answers to two questions, the quantifiers provide a way to combine the answers to a large (possibly infinite) number of questions. The boolean connectives allow us to ask different questions (formulas) in the same environment; the quantifiers allow us to ask the same question in many environments.[5] The existential quantifier forms the "OR" of the (possibly infinitely many)

[5]In $\forall vG$ or $\exists vG$ we may think of G as a subroutine body that is evaluated in an environment created by changing the value of v. This is a kind of call-by-value. Some authors define the quantifiers using something analogous to call-by-name (e.g., [Shoenfield 67]), but it comes out the same.

answers it receives; the universal quantifier forms the "AND" of the same set of answers.[6,7]

This concludes the definition of the language of first-order logic. It may seem somewhat rarified, since we have not written down or evaluated a single actual formula. We shall do that in the next section.

EXERCISES 5.6

1. Convert the following formulas to prefix form.

(a) $\underline{\forall x\, p(x) \supset \exists y q(y) \,\&\, {\sim} p(z) \vee q(z)}$

(b) $\underline{\forall x(p(x) \supset \forall y q(x,y) \supset \exists z\; q(z,x))}$

2. Show that "equivalent modulo v" is an equivalence relation.

3. Show that for any formula G and ivs v,

$$M(I, \underline{\forall} vG) = M(I, \underline{\sim}\ \underline{\exists}\, v \underline{\sim}\, G)$$

***4.** Extend Theorem 4.3.2 to the language of first-order logic.

***5.** Let $\mathfrak{F}$ be the set obtained by changing rule (1) of the definition of first-order formulas by replacing the word "terms" by "conditionals," and let M be the correspondingly changed evaluation function. Show that for any $G \in \mathfrak{F}$ there exists a first-order formula H such that for all $I, M(I,G) = M(I,H)$.

5.7 EXAMPLES IN FIRST-ORDER LOGIC

In this section we shall see how to evaluate first-order formulas over various data types.

Let $\mathcal{Q}$ be a data type with no functions and with a single, two-argument predicate R. Such a data type is still another picture of a directed graph (cf. Example 3.1.5). Here the universe is the set of nodes, and

$$R^{\mathcal{Q}}(x,y) = \text{TRUE} \quad \text{iff} \quad \text{there is an edge from } x \text{ to } y \text{ in } \mathcal{Q}.$$

[6]Other common notations for quantifiers include

$(\forall x)$	$(\exists x)$
$(\bigwedge x)$	$(\bigvee x)$
(Ax)	(Ex)
(x)	(Ex)

The $(\bigwedge x)$ and $(\bigvee x)$ notations suggest the infinite conjunction/disjunction picture. Some authors also write $p \wedge q$ for $p\&q$.

[7]One might wonder why one is allowed to take arbitrary *finite* boolean combinations, but one is allowed to take infinite conjunctions and disjunctions only of this very special kind. The easy answer is that one would like to stick with finite formulas (just as one would like to stick to finite-sized programs!), but fear not: There is an entire mathematical theory of infinite-length formulas.

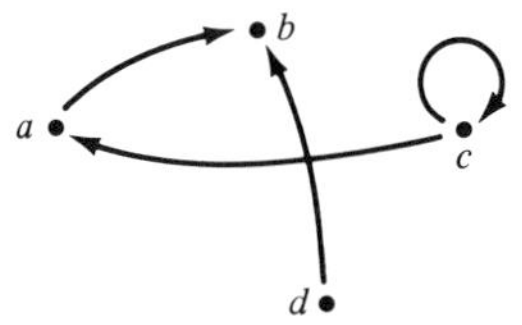

$$A=\{a,b,c,d\},$$

$$R^{\mathcal{A}}(x,y)=\begin{cases}\mathrm{T} & \text{if}\quad (x,y)=(a,b),\ (d,b),\ (c,a),\ \text{or}\ (c,c)\\ \mathrm{F} & \text{otherwise}\end{cases}$$

FIGURE 5.7.1 Graphical representation of $\mathcal{A}=\langle A,R^{\mathcal{A}}\rangle$.

In the next few examples we shall specify such a data type by drawing the directed graph (see Figure 5.7.1). Later on we shall work with more complicated data types.

EXAMPLE 1. Let $\mathcal{A}$ be the structure of Figure 5.7.1. Let $I=\{(\underline{\mathrm{x}},a),(\underline{\mathrm{y}},c)\}$.

$$\begin{aligned}M(I,\ \underline{\mathrm{R(x,y)}}\,)&=R^{\mathcal{A}}(M(I,\ \underline{\mathrm{x}}\,),M(I,\ \underline{\mathrm{y}}\,))\\ &=R^{\mathcal{A}}(I(\underline{\mathrm{x}}\,),I(\underline{\mathrm{y}}\,))\\ &=R^{\mathcal{A}}(a,c)\\ &=\mathrm{F}\end{aligned}$$

This shows the evaluation of atomic formulas. Normally, we shall skip the first two steps in the identity and write $M(I,\underline{\mathrm{R(x,y)}})=R^{\mathcal{A}}(a,c)=\mathrm{F}$.

EXAMPLE 2. With $\mathcal{A}$ and I as before,

$$\begin{aligned}M(I,\ \underline{\mathrm{R(x,x)\supset R(x,y)}}\,)&=\supset^{\mathcal{B}}(M(I,\ \underline{\mathrm{R(x,x)}}\,),M(I,\ \underline{\mathrm{R(x,y)}}\,))\\ &=\supset^{\mathcal{B}}(R^{\mathcal{A}}(a,a),R^{\mathcal{A}}(a,c))\\ &=\supset^{\mathcal{B}}(\mathrm{F},\mathrm{F})\\ &=\mathrm{T}\end{aligned}$$

EXAMPLE 3. Same $\mathcal{A}$, $J=\{(\underline{\mathrm{x}},b),\ (\underline{\mathrm{y}},c),\ (\underline{\mathrm{z}},a)\}$:

$$\begin{aligned}&M(J,\ \underline{\mathrm{R(x,y)\&R(y,z)\supset R(x,z)}}\,)\\ &\quad=\supset^{\mathcal{B}}(M(J,\ \underline{\mathrm{R(x,y)\&R(y,z)}}\,),M(J,\ \underline{\mathrm{R(x,z)}}\,))\\ &\quad=\supset^{\mathcal{B}}(\&^{\mathcal{B}}(M(J,\ \underline{\mathrm{R(x,y)}}\,),M(J,\ \underline{\mathrm{R(y,z)}}\,)),M(J,\ \underline{\mathrm{R(x,z)}}\,))\\ &\quad=\supset^{\mathcal{B}}(\&^{\mathcal{B}}(R^{\mathcal{A}}(b,c),R^{\mathcal{A}}(c,a)),R^{\mathcal{A}}(b,a))\\ &\quad=\supset^{\mathcal{B}}(\&^{\mathcal{B}}(\mathrm{F},\mathrm{T}),\mathrm{F})\\ &\quad=\supset^{\mathcal{B}}(\mathrm{F},\mathrm{F})\\ &\quad=\mathrm{T}\end{aligned}$$

So boolean combinations of atomic formulas are no problem to evaluate either.

EXAMPLE 4. Same $\mathcal{A}$ and I; let us try to evaluate $M(I, \underline{\exists x R(x,y)})$.[8] We shall have to evaluate $\underline{R(x,y)}$ in each environment I' such that $I' \underset{x}{\equiv} I$. Let us list the possibilities for I' and evaluate $\underline{R(x,y)}$ in each one.

I'		
$\underline{x}$	$\underline{y}$	$M(I', \underline{R(x,y)})$
a	c	F
b	c	F
c	c	T
d	c	F

When $I'(\underline{x}) = c$, $\underline{R(x,y)}$ is true, so $M(I, \underline{\exists x R(x,y)}) = T$.

EXAMPLE 5. $M(I, \underline{\forall x R(x,y)})$

We must use the same set of I' as in the previous example. Looking at the table, we see that not every line is T, so $M(I, \underline{\forall x R(x,y)}) = F$.

Another way of evaluating quantifiers uses a tree to keep track of the different environments. We write the starting environment at the root, and every time we alter an environment, we note the change along the branch. For the last pair of examples, we would have

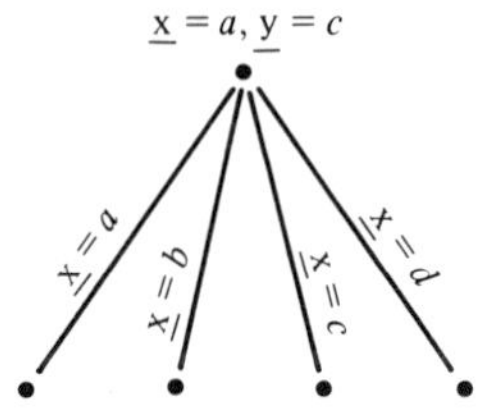

Every time we hit a quantifier, we split. For the moment, we shall ignore the problem of what to do with the boolean connectives (Exercise 5.7.7). Next let us label each node with the formula that is supposed to be evaluated at that node. Since all nodes at the same level will be evaluating the same formula, we can write the formula at the side. Last, we will draw a circular arc through the branches generated by an existential quantifier.[9] Figure 5.7.2 shows the general scheme.

[8]Before reading further, recite the formula aloud. What do you think the answer *should* be? As you grind through the calculation, make sure that it corresponds to your intuition. You should do this for each example.

[9]This is just to aid the eye in distinguishing the two types of nodes. This marking is also used in AND-OR trees, used in artificial intelligence [Nilsson 71].

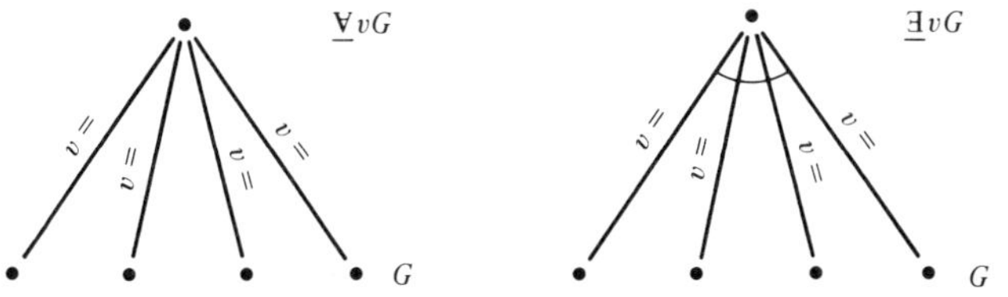

FIGURE 5.7.2 Interpretation trees for quantifiers.

At the leaves of this tree will be atomic formulas that we can evaluate, marking each leaf with T or F. We can then go back up through the tree, marking each node with T or F according to the appropriate quantifier rules. Figure 5.7.3 shows the completed trees for Examples 4 and 5.

Note that in Figure 5.7.3b, there was no need to compute the last three truth values, since a single F appearing at a son of a universal quantifier node will cause the father to be marked F. Similarly, a single T at a son of an existential node will cause the father to be marked T.[10]

This may seem like much wasted effort until we reach the case of nested quantifiers.

EXAMPLE 6. In $I=\{(\underline{x},(a)),(\underline{y},b)\}$ evaluate $\underline{\forall x \exists z(R(x,z)\&R(z,y))}$.

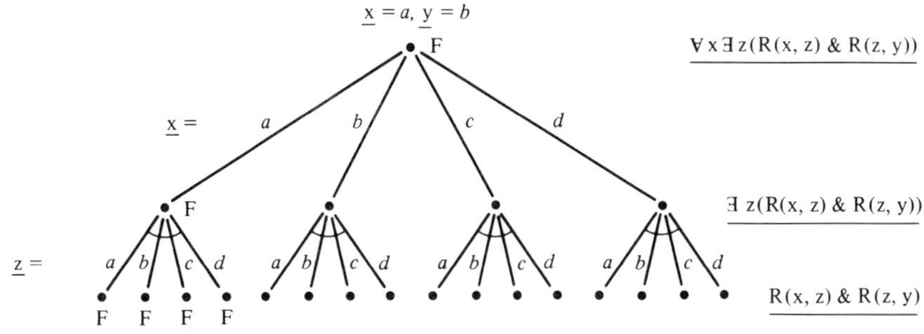

The answer is F.

$\underline{\exists z(R(x,z)\&R(z,y))}$ says there is a path of length 2 from x to y. So $\underline{\forall x\ \exists z(R(x,z)\&R(z,y))}$ says that for every x, there is a path of length 2 from x to y. Since we are working in an environment where $y=b$, we are asking the question: "For each node x, is there a path of length 2 from x to b?" And of course, setting $x=a$ quickly shows the answer is, "No."[11]

We evaluated individual variable symbols at the leaves by reading back up the appropriate branch. At each step some variable is "overridden."[12]

[10] Why? Justify this from the definition of M. This is called *pruning* the tree.

[11] Again, how is this "common-sense" analysis reflected in the tree?

[12] This is just like searching the association list in LISP or searching the static chain in ALGOL or PASCAL.

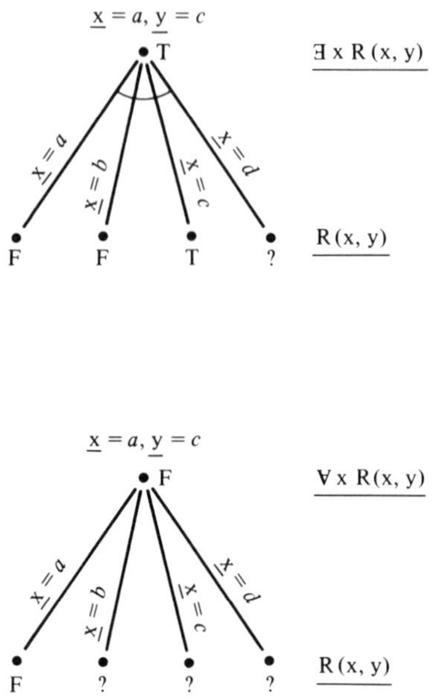

FIGURE 5.7.3

In Example 6, the initial value of x can never be referred to; nor could the initial value of z, if there were one. So if *every* variable symbol that appears in the formula is quantified over, the initial environment is completely ignored! (see Exercise 5.7.8). Such a formula is called a *sentence*.

EXAMPLE 7. In the structure of Figure 5.7.1, evaluate

$$\underline{\forall x\ \forall y(R(x,y) \supset \sim R(y,x))}$$

Since every variable is quantified over, we need not specify an initial environment.

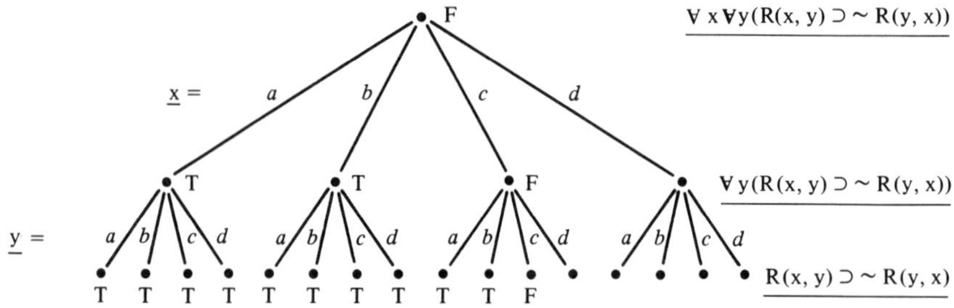

At $x=c$, $y=c$, the formula comes out false. This is reasonable, since it says that if there is an arrow from x to y, then there is no arrow from y to x; this is false when $x=y=c$.

EXAMPLE 8. $\forall x\, \exists y(R(x,y) \vee R(y,x))$

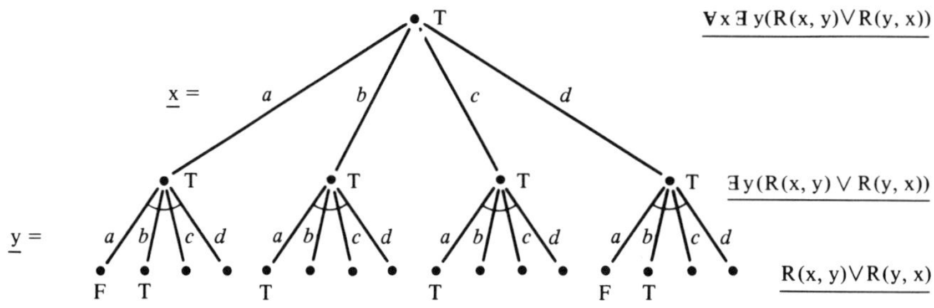

$R(x,y) \vee R(y,x)$ says that there is an arrow from x to y or from y to x; $\exists y(R(x,y) \vee R(y,x))$ says that there is at least one arrow into or out of x; so $\forall x\, \exists y(R(x,y) \vee R(y,x))$ says that there is an arrow either into or out of every node—which is true.

EXAMPLE 9. $\exists y\, \forall x(R(x,y) \vee R(y,x))$

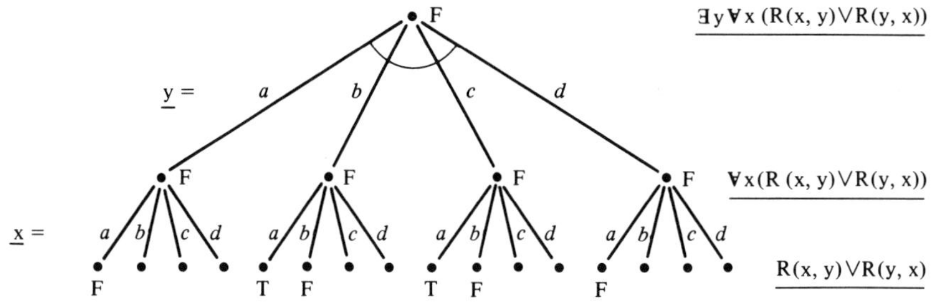

$\forall x(R(x,y) \vee R(y,x))$ says that, given y, there is an arrow connecting y with each node. $\exists y\, \forall x(R(x,y) \vee R(y,x))$ says there exists a node with this property. We may go through the graph and verify that no node has the property; and indeed the formula evaluates to false.

Notice that the sole difference between Examples 8 and 9 was the *order* of the quantifiers. $\forall x \exists y G$ says that for each x there exist a y that makes G true—possibly a different y for each x. (Table 5.7.1 shows one choice of y for each x in Example 8.) $\exists y \forall x G$ says that there is a *single* y that works for all choices of x. $\forall$ and $\exists$ do not commute![13]

[13]Formulas with many alternations of quantifiers are, of course, more difficult to describe intuitively. Luckily, such formulas do not arise often.

TABLE 5.7.1

x	y
a	b
b	a
c	a
d	b

The methods we have described in this section extend to arbitrarily complex formulas (see Exercise 5.7.7). Let us look at one last example to show how to deal with data types that have functions.

Let $\mathcal{C}$ be the data type $\mathcal{Q}$ with equality and one-argument function $f^{\mathcal{C}}$ added, where $f^{\mathcal{C}}$ is defined by:

$$f^{\mathcal{C}} = \{(a,b),(b,c),(c,d),(d,a)\}$$

EXAMPLE 10. $\underline{\forall x\ \exists y(f(y)=x)}$

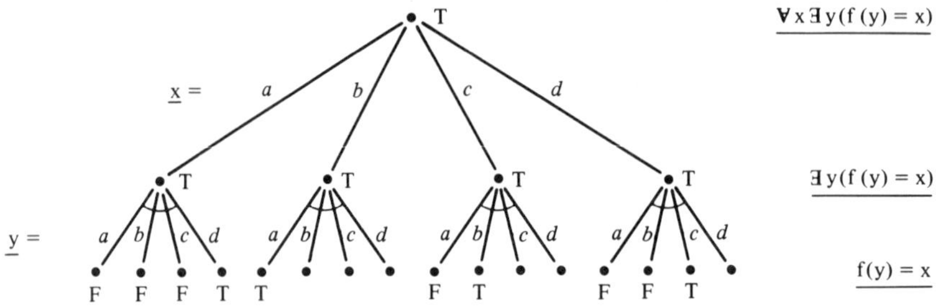

The formula says that for every x there is a y such that $f(y)=x$ (i.e., the function f is "onto" or "surjective"), and indeed it is.

EXERCISES 5.7

Let $\mathcal{G} = \langle\{a,b,c\}, R^{\mathcal{G}}, E^{\mathcal{G}}\rangle$ where $R^{\mathcal{G}}$ and $E^{\mathcal{G}}$ are two-place predicates:

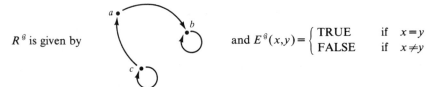

$R^{\mathcal{G}}$ is given by

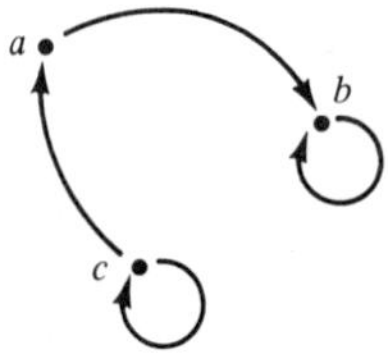

and $E^{\mathcal{G}}(x,y) = \begin{cases} \text{TRUE} & \text{if } x=y \\ \text{FALSE} & \text{if } x \neq y \end{cases}$

For each of the following formulas, paraphrase the formula in English and determine whether or not the formula is valid in $\mathcal{G}$:

1. $\underline{\forall x\ R(x,x)}$

2. $\forall x\, \exists y\, R(x,y)$

3. $\forall x\, \forall y\, \forall z[R(x,z) \& R(y,z) \supset E(x,y)]$

Let $\mathcal{K}=\langle\{0,1\},f^{\mathcal{K}},E^{\mathcal{K}}\rangle$ where $f^{\mathcal{K}}$ is a two-place function given by

$f^{\mathcal{K}}$	0	1
0	0	1
1	1	0

and $E^{\mathcal{K}}$ is the two-place predicate given by

$$E^{\mathcal{K}}(x,y)=\begin{cases}\text{TRUE} & \text{if } x=y\\ \text{FALSE} & \text{if } x\neq y\end{cases}$$

Do the same thing for the following formulas:

4. $\exists x\, \forall y[E(f(x,y),y) \& E(f(y,x),y)]$

5. $\forall x\, \forall y[E(f(x,y),f(y,x))]$

6. If $\mathcal{A}$ is any data type, show that if $M(I,\exists y \forall x G)=\text{TRUE}$, then $M(I, \forall x \exists y G)=\text{TRUE}$.

7. Extend the evaluation methods of this section to deal with *any* first-order formula in a finite data type.

8. We say a variable v appears *free* in G if it appears in the formula and is not within the scope of any quantifier $\forall v$ or $\exists v$. Prove, using the definition of M, that if I and I' agree on the free variables of G, then $M(I,G)=M(I',G)$.

5.8 DEGREES OF TRUTH

In the last two sections, we considered the problem of determining the truth or falsity of a formula in a data type at a particular interpretation. This still does not give us the notion of a "fact about a data type" because we still have that pesky interpretation or environment parameter to worry about. In this section we shall show how to eliminate that parameter. We shall also give examples of how common facts about data types may be expressed using the language of first-order logic.

Definition. If $\mathcal{A}$ is a data type, we say $\mathcal{A}$ *has an equality predicate* if $\mathcal{A}$ has a two-place predicate $=^{\mathcal{A}}$ defined by

$$=^{\mathcal{A}}(x,y)=\begin{cases}\text{TRUE} & \text{if } x=y\\ \text{FALSE} & \text{if } x\neq y\end{cases}$$

Note that any data type can be converted into a data type with an equality predicate by adding a new two-place predicate $=^{\mathcal{A}}$. Henceforth we assume all of our data types have equality predicates. We will use infix notation for equality: e.g., $x=y$.

Definition. If G is a formula and $\mathfrak{A}$ is a data type with an equality predicate, we say G is *valid in* $\mathfrak{A}$ (write $\mathfrak{A}\models G$)[14] iff for every I: IVS$\rightarrow A$, $M(I,G)=\mathrm{T}$.

G is valid in $\mathfrak{A}$ iff it is true at every environment over $\mathfrak{A}$. Thus, $\mathfrak{A}\models G$ is our notion of G being a fact in the data type $\mathfrak{A}$. We interpret $\mathfrak{A}\models G$ as "$\mathfrak{A}$ makes G true" or "$\mathfrak{A}$ is a model for G" or "$\mathfrak{A}$ satisfies G."

Now we can give some examples of facts in various data types.

Let $R^{\mathfrak{A}}$ be a two-argument predicate in a data type $\mathfrak{A}$ with an equality predicate. Then

1. $R^{\mathfrak{A}}$ is transitive iff $\mathfrak{A}\models\underline{(\forall x)(\forall y)(\forall z)[R(x,y)\&R(y,z)\supset R(x,z)]}$
2. $R^{\mathfrak{A}}$ is reflexive iff $\mathfrak{A}\models\underline{(\forall x)[R(x,x)]}$
3. $R^{\mathfrak{A}}$ is symmetric iff $\mathfrak{A}\models\underline{(\forall x)(\forall y)[R(x,y)\supset R(y,x)]}$
4. $R^{\mathfrak{A}}$ is a function iff
$\mathfrak{A}\models\underline{(\forall x)(\exists y)[R(x,y)]\&(\forall x)(\forall y)(\forall z)[R(x,y)\&R(x,z)\supset y=z]}$

Let $f^{\mathfrak{A}}$ be a one-argument function in a data type $\mathfrak{A}$. Then

5. f is onto iff $\mathfrak{A}\models\underline{(\forall y)(\exists x)[f(x)=y]}$

Let $\mathfrak{Z}$ be the data type of integers (Example 3.1.1). Some typical sentences which are valid in $\mathfrak{Z}$ are

i. $\underline{(\forall x)(\forall y)[x+y=y+x]}$ (addition is commutative)
ii. $\underline{(\forall x)(\forall y)(\forall z)[(x+y)+z=x+(y+z)]}$ (addition is associative)
iii. $\underline{(\forall x)(\forall y)(\exists z)[x+z=y]}$ (cancellation)
iv. $\underline{(\forall x)(\forall y)[(x+y)-y=x]}$
v. $\underline{(\forall x)(\forall y)(\forall z)[(x+y)-z=(x-z)+y]}$
vi. $\underline{(\forall x)[x<(x+1)]}$
vii. $\underline{(\forall x)(\forall y)(\forall z)[x<y\&y<z\supset x<z]}$
viii. $\underline{(\forall x)(\forall y)[(x<y)\vee(y<x)\vee(y=x)]}$

Here we have used typical infix notation for the arithmetic operations.

[14]"$\models$" is read "double turnstile."

Let $\mathbb{S}$ be the data type of stacks, with equality added. Typical sentences valid in $\mathbb{S}$ are

i. $\underline{(\forall x)[\text{is0?}(\text{push0}(x))]}$

ii. $\underline{(\forall x)[\text{pop}(\text{push0}(x)) = x]}$

iii. $\underline{(\forall x)[\text{is0?}(x) \supset \text{push0}(\text{pop}(x)) = x]}$

One might also ask: Are there formulas which are true in *every* data type with equality? The answer is yes:

Definition. If G is a formula, then G is *valid* iff for every data type $\mathfrak{A}$ with equality, $\mathfrak{A} \models G$.

G is valid iff it is valid in *every* data type. Thus a valid formula is not so much a fact about a particular data type as a fact about the way functions and predicates behave in general.

Some examples of valid formulas are

$\underline{\forall x \forall y\, p(x,y) \supset \forall y \forall x\, p(x,y)}$

$\underline{\exists x \forall y\, p(x,y) \supset \forall y \exists x\, p(x,y)}$

$\underline{\forall x \forall y [x = y \supset y = x]}$

$\underline{\exists x (x = x)}$

$\underline{\forall x (p(x) \,\&\, q(x)) \supset (\forall x\, p(x)) \& (\forall x\, q(x))}$

$\underline{\forall x (p(x) \supset q(x)) \supset (\forall x\, p(x) \supset \forall x\, q(x))}$

Unfortunately, we cannot use the methods of Section 7 to demonstrate any of these last examples, since both $\mathbb{Z}$ and $\mathbb{S}$ have infinite carriers; we would have to search trees in which each node has infinitely many sons! Consequently, we need some better methods for actually computing truth values. If we are interested in validity, the situation is even worse: we must also search over all data types! We will take up this problem in the next section.

EXERCISES 5.8

Translate the following properties into first-order formulas:

1. R (a two-place predicate) is an injective function.

2. R is an equivalence relation.

3. f (a two-place function) is associative.

4. c (a constant) is an identity for f.

Determine the validity of each of the following formulas. If it is not valid, give a data type and an interpretation in which it is false. If it is valid, give an argument for its validity.

5. $\underline{p(x) \supset \exists x\, p(x)}$

6. $\underline{\exists x\, p(x) \supset p(x)}$

7. $\underline{\exists x\, p(x) \,\&\, \exists x\, q(x) \supset \exists x(p(x) \,\&\, q(x))}$

8. $\underline{\exists x\, p(x) \vee \exists x\, q(x) \supset \exists x(p(x) \vee q(x))}$

9. $\underline{\forall x\, \exists y\, p(x,y) \supset \exists y\, p(z,y)}$

10. Show that the relation $A=\{1\}$, $R=\varnothing$ is transitive.

5.9 PROVABILITY

In previous sections, we had an inductive definition of the function M, which we then used to define the set of valid formulas. As we saw, it is generally impossible to determine the validity of a formula directly from the definition of validity. In this section, we shall try a different approach: We shall attempt to give a direct inductive definition of the set of valid wffs.

An inductive definition of a set S consists, of course, of a base set of formulas and some rules, roughly of the form "if $x \in S$, then $f(x) \in S$." In logic, the formulas in the base set are called *axioms* and the rules are called *rules of inference*. A derivation, showing how a particular formula gets into the set (like the computations in Section 2.4, or parsing as in Section 3.4) is called a *formal proof*; a formula that is in the set is called *provable*. A definition in this style is called a *formal system*.

Since the set of valid formulas is rather complicated, it should not come as a shock that the inductive definition is fairly complicated. In fact, in order to keep things reasonable, we shall fall back to dealing with propositional logic.[15]

Our new, more modest aim is to give an inductive definition of the set of all tautologies. In order to do this, let FW denote the set of all finite sets of wffs, and define a relation $\models$ on FW as follows:

[15]Actually, the extension of system G to first-order logic is not much different from the system G given here, but it requires a few technical notions that are beyond the scope of this course. System G as given here is based on [Kleene 52, §7]. A good discussion of a similar system, including the extension to first-order logic, is given in [Manna 74, Section 2-2.] For more on formal proofs, see [Kalish and Montague 64] or [Hofstadter 79].

Definition. If $\Delta \in FW$ and $\Gamma \in FW$, $(\Delta, \Gamma) \in \models$ iff every interpretation that makes all of the formulas in Δ true makes at least one of the formulas in Γ true.[16]

Thus

$$(\{p, p \supset q \vee \sim r\}, \{q, \sim r\}) \in \models$$

but

$$(\{p \vee q, \sim q \supset r\}, \{\sim p \supset \sim r\}) \notin \models,$$

Since $I = \{(p, \mathrm{F}), (q, \mathrm{T}), (r, \mathrm{T})\}$ makes both $p \vee q$ and $\sim q \supset r$ true and $\sim p \supset \sim r$ false.

In keeping with standard notation, we write $\Delta \models \Gamma$ when $(\Delta, \Gamma) \in \models$, and we usually delete the braces when listing the elements of the sets Δ and Γ. Thus we write

$$p, p \supset q \vee \sim r \models q, \sim r$$

(Thus "," gets precedence lower than "$\supset$", and "$\models$" gets lowest precedence of all: these symbols are not parts of wffs).

In particular, a wff A is a tautology iff $\varnothing \models \{A\}$.

This $\models$, however, is not the inductive definition we are seeking. It still involves "every interpretation...." We next define inductively a relation $\vdash$ on FW and prove, using the fundamental theorem on induction (Section 2.3) that $\vdash = \models$. Then we will have shown that A is a tautology iff $\varnothing \vdash \{A\}$.

To make our definition of $\vdash$ easier to read, we shall write $\vdash$ in the infix position and delete braces, as we did for $\models$. We shall also write Δ, A for $\Delta \cup \{A\}$. We write

$$\frac{\Delta \vdash \Gamma}{\Delta' \vdash \Gamma'}$$

for "if $(\Delta, \Gamma) \in \vdash$, then $(\Delta', \Gamma') \in \vdash$," and

$$\frac{\Delta_1 \vdash \Gamma_1 \quad \Delta_2 \vdash \Gamma_2}{\Delta_3 \vdash \Gamma_3}$$

for "if $(\Delta_1, \Gamma_1) \in \vdash$ and $(\Delta_2, \Gamma_2) \in \vdash$, then $(\Delta_3, \Gamma_3) \in \vdash$."

It will be helpful to have names for the rules. We name a rule by placing its name to the right of the line:

$$\frac{\Delta_1 \vdash \Gamma_1 \quad \Delta_2 \vdash \Gamma_2}{\Delta_3 \vdash \Gamma_3}\ \xi$$

names this rule ξ.

[16] In the terminology of Exercise 5.2.3, the conjunction of the wffs in Δ logically implies the *dis*junction of the wffs in Γ. This connection is the reason we use the "double turnstile" symbol for both. Why the disjunction of the wffs in Γ? Because it works: the inductions go through correctly. The cleverness of which good mathematics is created is rarely so obvious as it is here.

Table 5.9.1 shows the inductive definition of $\vdash$. We will call this definition *System G*.[17]

How are we to use this system? Presume we are given a pair (Δ, Γ) and asked to determine whether it is a member of $\vdash$. We pick a rule that might have introduced (Δ, Γ), say via

$$\frac{\Delta' \vdash \Gamma'}{\Delta \vdash \Gamma}$$

We then work on (Δ', Γ') in a similar manner, until we reach an axiom or until we can proceed no further. This is best done as a tree, in which we start with (Δ, Γ) as the root and proceed toward the leaves. Let us do an example:

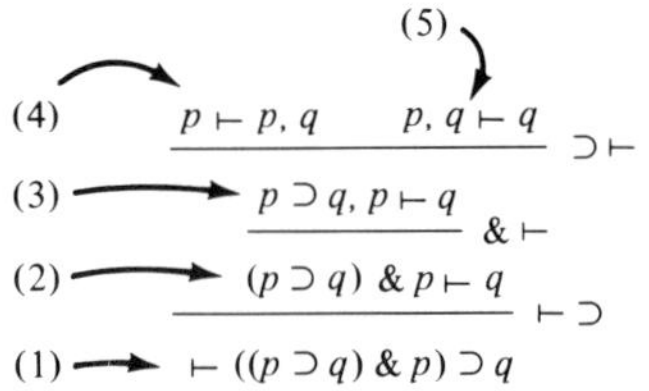

Here we started out with $(\varnothing, \{((p \supset q) \& p) \supset q\})$. The only way this could be in $\vdash$ is through an instance of rule $\vdash\supset$ (4b), with $\Delta = \Gamma = \varnothing$, $A = (p \supset q) \& p$, and $B = q$. This leaves us with (2): $(\{(p \supset q) \& p\}, \{q\})$. The only way that (2) can get into G is by rule $\&\vdash$ (3a), from $(\{p \supset q, p\}, \{q\})$. So we write that down (3). Again, the only way that (3) can get in is via rule $\supset\vdash$ (4*a*), giving (4) and (5). Now, (4) has p on both sides of the turnstile, and (5) has q on both sides of the turnstile, so these are both axioms, and therefore in G. We have now constructed a tree that shows how, starting with $p \vdash p, q$ and $p, q \vdash q$, we can apply the rules to show that $\vdash ((p \supset q) \& p) \supset q$ is a member of G.

Let us look a little more closely at the rules. Notice that they come in pairs. For each connective there is a pair of rules: one rule about the connective to the left of the "$\vdash$" and one rule about the connective to the right of the "$\vdash$." This is the motivation for the rule names, that appear to the right of each rule in Table 5.9.1 ("$\vee\vdash$," for example, may be read as "OR on the left"). Each rule tells how a wff with that main connective can appear on that side of the $\vdash$. So, to show $(\Delta, \Gamma) \in \vdash$, pick a wff in Δ or Γ (not necessarily one adjacent to the "$\vdash$"), find its main connective, and apply the appropriate rule "backwards." Thus we start at the root and work towards the leaves; we start at the conclusion and work towards the

[17] We name it system *G* after Gerhard Gentzen, who originated this type of formal system.

TABLE 5.9.1 System *G*

(1)	If $\Delta\cap\Gamma\neq\varnothing$, then $\Delta\vdash\Gamma$		(axioms)
(2)	$\dfrac{\Delta,A\vdash\Gamma \quad \Delta,B\vdash\Gamma}{\Delta,A\vee B\vdash\Gamma}\ \vee\vdash$	$\dfrac{\Delta\vdash A,B,\Gamma}{\Delta\vdash A\vee B,\Gamma}\ \vdash\vee$	(rules)
(3)	$\dfrac{\Delta,A,B\vdash\Gamma}{\Delta,A\,\&\,B\vdash\Gamma}\ \&\vdash$	$\dfrac{\Delta\vdash A,\Gamma \quad \Delta\vdash B,\Gamma}{\Delta\vdash A\,\&\,B,\Gamma}\ \vdash\&$	(rules)
(4)	$\dfrac{\Delta\vdash\Gamma,A \quad \Delta,B\vdash\Gamma}{\Delta,A\supset B\vdash\Gamma}\ \supset\vdash$	$\dfrac{\Delta,A\vdash B,\Gamma}{\Delta\vdash A\supset B,\Gamma}\ \vdash\supset$	(rules)
(5)	$\dfrac{\Delta\vdash A,\Gamma}{\Delta,\sim A\vdash\Gamma}\ \sim\vdash$	$\dfrac{\Delta,A\vdash\Gamma}{\Delta\vdash\sim A,\Gamma}\ \vdash\sim$	(rules)
	(a)	(b)	

axioms.[18] As we apply each rule from bottom to top, the main connective disappears. Hence the process of applying the rules backwards eventually halts.

Let us try another example:

$$\dfrac{\dfrac{\dfrac{p\vdash p,q}{p,\sim p\vdash q}\ \sim\vdash \qquad q,\sim p\vdash q}{p\vee q,\sim p\vdash q}\ \vee\vdash}{(p\vee q)\,\&\sim p\vdash q}\ \&\vdash$$

Here, at the second step, we could have applied either $\vee\vdash$ or $\sim\vdash$ to $p\vee q,\sim p\vdash q$. We chose $\vee\vdash$, but $\sim\vdash$ would also have worked:

$$\dfrac{\dfrac{\dfrac{p\vdash p,q \quad q\vdash p,q}{p\vee q\vdash p,q}\ \vee\vdash}{p\vee q,\sim p\vdash q}\ \sim\vdash}{(p\vee q)\,\&\sim p\vdash q}\ \&\vdash$$

One of the most pleasant properties of system G is that, if $(\Delta,\Gamma)\in\vdash$, no matter how you choose to build the tree you are guaranteed to succeed in building the tree. Let us try one more:

$$\dfrac{p\supset q,p,r\vdash r \qquad \dfrac{\dfrac{p,r,q\vdash p \qquad \overset{?}{q},\overset{?}{p},\overset{?}{r}\overset{?}{\vdash}}{p\supset q,p,r,q\vdash}\ \supset\vdash}{p\supset q,p,r\vdash\sim q}\ \vdash\sim}{p\supset q,p,r\vdash r\,\&\sim q}\ \vdash\&$$

[18]Again, this is similar to a "functional form" calculation (Section 2.4), in which the rules guide us in deducing which of the axioms or basis conditions we need. Unlike the functional form calculation, we do not have to worry about recombining the values.

Here we applied the rules until we got $p \supset q, p, r \vdash r$, which is an axiom (even though it has a wff with a connective), $p, q, r \vdash p$, which is an axiom, and $q, p, r \vdash$, which is not an axiom and which cannot be broken down any further. Now, if $p \supset q, p, r \vdash r \& \sim q$ were a member of $\vdash$, then *any* choice of how to build the tree would have reached axioms at every leaf. So there is no point to trying another derivation: $p \supset q, p, r \vdash r \& \sim q$ is *not* in G. As soon as we reach a node with $\Delta \vdash \Gamma$, where Δ and Γ have no connectives and $\Delta \cap \Gamma = \varnothing$, we can quit: the object at the root of the tree is not in G. Of course, we have only *claimed* that G has this pleasant property; we shall eventually have to prove it.[19]

Note also that whenever $\Delta \cap \Gamma \neq \varnothing$, then $\Delta \vdash \Gamma$, even if Δ and Γ still have connectives. Thus

$$\dfrac{\dfrac{p \supset ((q \supset r) \supset s) \vdash p \supset ((q \supset r) \supset s)}{\vdash (p \supset ((q \supset r) \supset s), \sim (p \supset ((q \supset r) \supset s)} \vdash\sim}{\vdash (p \supset ((q \supset r) \supset s)) \vee \sim (p \supset ((q \supset r) \supset s))} \vdash\vee$$

is a correct derivation in system G.

Now that we have gained some familiarity with the mechanical manipulations in System G, we can start making the connections between "provability" in G and "truth values" in propositional logic.

Theorem 5.9.1. *If* $\Delta \vdash \Gamma$, *then* $\Delta \vDash \Gamma$.

PROOF. We will show (1) that if (Δ, Γ) is an axiom of system G, then $\Delta \vDash \Gamma$, and (2) $\vDash$ is closed under the rules of system G. Hence, by the fundamental theorem on induction (Theorem 2.3.1), $\vdash \subseteq \vDash$. Our proof will thus have a base step and one step for each of the rules of system G. We shall do a few cases here and leave the others as exercises.

Base Step: let $\Delta \vdash \Gamma$ be an axiom. Let $C \in \Delta \cap \Gamma$. Every interpretation I which makes all of the wffs in Δ true makes C true, so I makes one of the wffs in Γ true also (in particular, it makes C true).

Rule $\&\vdash$*:* Assume $\Delta, A, B \vDash \Gamma$. We want to show $\Delta, A \& B \vDash \Gamma$. So let I make all of the formulas in $\Delta \cup \{A \& B\}$ true. In particular, I then makes A and B both true. So I makes every formula in $\Delta \cup \{A, B\}$ true. Since $\Delta, A, B \vDash \Gamma$, I makes some formula in Γ true. Hence $\Delta, A \& B \vDash \Gamma$.

Rule $\vdash\&$*:* Assume $\Delta \vDash A, \Gamma$ and $\Delta \vDash B, \Gamma$. We must show $\Delta \vDash A \& B, \Gamma$. Let I make all of the formulas in Δ true. We must show I makes some formula in Γ true or I makes $A \& B$ true. Since $\Delta \vDash A, \Gamma$ and $\Delta \vDash B, \Gamma$, either I makes some formula in Γ true or it must make both A true and B true. Hence I makes some formula in $\Gamma \cup \{A \& B\}$ true.

[19] Indeed, it should not be at all obvious that G has this property. That is precisely why proofs (in this case, proofs about proofs!) come in handy.

Rule $\vee\vdash$: Assume $\Delta, A \models \Gamma$ and $\Delta, B \models \Gamma$. If interpretation I makes every wff in $\Delta, A \vee B$ true, then it either makes A true or B true. If it makes A true, then it makes all of the wffs in Δ, A true and so it makes some wff in Γ true. If it makes B true, then it makes all of the wffs in Δ, B true, so it makes some wff in Γ true. So either way I makes some wff in Γ true. So $\Delta, A \vee B \models \Gamma$.

Rule $\vdash\sim$: Assume $\Delta \models A, \Gamma$. Let I make every wff in $\Delta, \sim A$ true. So it makes every wff in Δ true and it makes A false. Since I makes every wff in Δ true, it must make some wff in A, Γ true. It cannot make A true (since it already makes A false), so it must make some wff from Γ true. So $\Delta, \sim A \models \Gamma$.

The other cases are similar and are left as an exercise. □

Of course, we also have to characterize the things that are *not* in G. To do this, we shall temporarily consider what happens if we allow *any* (Δ, Γ) with no connectives to appear as a leaf of the tree.

Theorem 5.9.2. *If there is any derivation of* (Δ, Γ) *using the rules in* G *in which some leaf contains no connectives but is not an axiom, then* $(\Delta, \Gamma) \notin \models$, *that is, there is an interpretation* I *that makes every member of* Δ *true and every member of* Γ *false.*

PROOF

Base Step: If (Δ, Γ) contains no connectives and $\Delta \cap \Gamma = \varnothing$, then make I true on all the propositional variables in Δ and false on all the propositional variables in Γ.

Induction Step: Again we have one case for each rule.

Rule $\&\vdash$: If I makes Δ, A, B all true and Γ all false, then I makes $\Delta, A \& B$ all true and Γ all false.

Rule $\vdash\&$: Assume the nonaxiom leaf is in the derivation of $(\Delta, \{A\} \cup \Gamma)$. Then there is some I that makes Δ all true and A, Γ all false. Then I makes Δ all true and $A \& B, \Gamma$ all false.

Rule $\vee\vdash$: Assume that there is an I that makes Δ, A all true and Γ all false. Then it makes $\Delta, A \vee B$ all true as well.

Rule $\vdash\sim$: Assume we have an I that makes Δ all true and A, Γ all false. Then it makes $\Delta, \sim A$ all true (since it makes A false), and Γ all false.

The other cases are similar. □

Corollary 5.9.3. *If there is any derivation of* (Δ, Γ) *using the rules of* G *in which some leaf contains no connectives and is not an axiom, then* $(\Delta, \Gamma) \notin \vdash$.

PROOF. If there is such a derivation, then by Theorem 5.9.2, $(\Delta,\Gamma)\notin\models$. But by Theorem 5.9.1, $\vdash\subseteq\models$, so $(\Delta,\Gamma)\notin\models$ implies $(\Delta,\Gamma)\notin\vdash$. □

Corollary 5.9.3 makes precise the "pleasant property" we discussed earlier in this section.[20]

Corollary 5.9.4. $(\Delta,\Gamma)\in\vdash$ *iff* $(\Delta,\Gamma)\in\models$.

PROOF. Theorem 5.9.1 showed $\vdash\subseteq\models$; it remains to show that $\models\subseteq\vdash$. Assume $\Delta\models\Gamma$. Construct a derivation of (Δ,Γ) using the rules of system G until every leaf is either an axiom or a nonaxiom with no connectives. This process will always halt, no matter how the wffs are chosen for decomposition, since you always lose one connective at each step along a branch. But no leaf in the resulting tree can be a nonaxiom, since then, by Theorem 5.9.2, (Δ,Γ) would not be a member of $\models$. Hence the process yields a derivation of (Δ,Γ) from the axioms of system G, and $(\Delta,\Gamma)\in\vdash$. □

Corollary 5.9.5. *$\varnothing\vdash A$ iff A is a tautology.* □

Corollary 5.9.5 proves that G has the property we originally wanted to have: $(\varnothing,\{A\})\in G$ iff A is a tautology. We can build a similar system for first-order logic, in which only valid formulas are generated, but the counterpart of Theorem 5.9.2 fails. In fact, the situation is even worse than that: It can be shown that there is no algorithm that, given a first-order formula, will always halt and tell whether or not the formula is valid [Manna 74, Theorem 2-6]. So some kind of intelligence is *required* to examine formulas for validity or for validity in a data type.[21]

In the next chapter we shall assume that simple facts about a data structure may be affirmed or denied "by inspection" and take up again the problem of proving the correctness of algorithms.

EXERCISES 5.9

Using the system G, verify:

1. $p\supset q, p\supset r\vdash p\supset r$.

2. $\sim p\vdash(\sim q\supset p)\supset q$.

[20]In light of Theorem 5.9.2, we can regard the construction of the tree as a search for a *refutation* of $\Delta\models\Gamma$, i.e., a search for an I that makes all of Δ true and all of Γ false. Either the search finds such an I, or it fails, finding a derivation of $\Delta\vdash\Gamma$. This is called *Wang's algorithm* [McCarthy et al. 65, Chapter 8]; for more on such procedures, see [Nilsson 71].

[21]This is *not* meant to imply that human intelligence is somehow mystically different from machine intelligence. That is a philosophical debate, which generates much heat but little light. However, the theorem surely seems to apply to algorithms implemented in "brainware", since there are simple formulas whose validity (say in $\mathfrak{N}$) remains unknown—e.g., Fermat's last theorem, Goldbach's conjecture. For more on this issue, see [Hofstadter 79].

3. $\vdash (p \supset \sim\sim p) \& (\sim\sim p \supset p)$.

4. $p \& \sim p \vdash q$.

5. $\vdash (\sim(p \& q) \supset (\sim p \vee \sim q)) \& ((\sim p \vee \sim q) \supset \sim(p \& q))$.

6. Complete the proof of Theorem 5.9.1.

7. Complete the proof of Theorem 5.9.2.

6 Proving Assertions About Programs

In Chapter 3, we studied a language for the precise statement of algorithms. At the end of that chapter, we proved the correctness of some simple algorithms, and we saw that the correctness of each algorithm depended on certain facts about the data type in which we were computing. In Chapter 5, we studied the language of logic, which was a language for the precise statement of such facts. In this chapter, we shall study the language of *assertions*, which will be a language for the precise statement of assertions about the correctness of programs. We shall introduce a formal system called *system H*, which will help us analyze assertions to determine what properties of the data type are crucial to the correctness of a program. If these properties, called *verification conditions*, are true in the data type, then we may conclude that the program is correct.

We could, if we wished, carry out this plan using the language of expressions from Chapter 3 (see, e.g., [Morris and Wegbreit 77]). We prefer, however, to use a language closer to those typically used for discussing the verification of programs. We therefore use the language of *statements* to express the programs that we shall prove correct. We have already used this language, without fanfare, in Section 1.5; it is probably similar to languages with which you are familiar.

The techniques of this chapter are crucial in modern software engineering, where they form the topic called *program verification*. Program verification is one of the practices advocated by the *structured programming* movement. We shall see how the ideas of program verification allow the programmer to prove (at least informally) the correctness of his program *as he writes it*. Once a correct program is obtained, the same techniques often show how to improve the performance of a program.

This programming paradigm can decrease debugging time to close to nil. F. T. Baker and Harlan Mills (of IBM), two leaders of the structured programming movement in industry, report that on some programming projects, the use of structured programming techniques has decreased error rates to under one bug per *thousand* lines of code [Baker 72]! This was achieved by a combination of the mathematical techniques of this chapter and some new project management techniques.

For more on "structured programming," see the special issue of *Computing Surveys* for December, 1974, which contains several articles on this topic. For some indication of the continuing interest in these topics, see [ICRS 75].

6.1 THE LANGUAGE OF STATEMENTS

In this section we shall define the language of *statements*, which will be the minilanguage for expressing programs for this chapter. The language of statements is very close to FORTRAN, PASCAL, PL/I, etc., in that the fundamental operation is the *assignment* of a value to a variable.[1] Assignments are combined by sequencing, conditionals, and loops. Our presentation will follow the familiar pattern: we shall define by induction the strings in the language and then we shall define a meaning function, also by induction.

As usual assume we are working in a data type $\mathcal{A}$.

Definition. If $\mathcal{A}$ is a data type, the set of *statements over* $\mathcal{A}$ is the set of strings defined as follows:

(i) if $v \in \mathrm{IVS}$ and t is a term over $\mathcal{A}$ (see Section 3.3), then $v \underline{\leftarrow} t$ is a statement
(ii) if S_1 and S_2 are statements, then $\underline{\text{begin}}\ S_1 \underline{;}\ S_2\ \underline{\text{end}}$ is a statement
(iii) if B is a formula over $\mathcal{A}$ with no quantifiers (see Section 5.6), and S_1 and S_2 are statements, then $\underline{\text{if}}\ B\ \underline{\text{then}}\ S_1\ \underline{\text{else}}\ S_2$ is a statement
(iv) if B is a formula over $\mathcal{A}$ with no quantifiers and S is a statement, then $\underline{\text{while}}\ B\ \underline{\text{do}}\ S$ is a statement
(v) nothing else

Figure 6.1.1 shows the names and intended flowcharts for each of these classes of statements.[2] The interesting part, as usual, is defining the

[1]Languages that are organized around assignment are called *imperative* languages. Languages, such as the language of expressions, that are organized around the application of functions to arguments, are called *applicative* languages [Landin 66]. Most real programming languages have both applicative and imperative features, and each organization seems to allow insights not offered by the other. It is therefore important to be familiar with both.

[2]One aspect of structured programming is the elimination of unconstrained go-to's in favor of these three constructions. Note that each flowchart has precisely one entrance and one exit, and that each construction corresponds to a natural way of dividing a problem into

evaluation function M. Statements operate by changing the values of variables, so the key concept is the map from variable names to their values—that is, the environment.[3] So we shall make M a partial function that takes a statement S and an environment I (the starting environment) and that returns an environment I' (the environment after execution of the statement). Table 6.1.1 gives the definition of M. We have written the definition in functional form (see Section 2.4), but remember that this is just convenient notation for a relational form definition; thus we occasionally need to write $(I,S,I')\in M$ instead of $M(I,S)=I'$. Similarly, for some I,S, there is no I' such that $(I,S,I')\in M$. In that case $M(I,S)$ is undefined; this corresponds to a nonterminating computation. We shall not attempt to do calculations with this definition, as we did in Chapter 3, but you should understand the following things about the definition:

1. what each statement does;
2. how the line in the definition of M corresponds to the flowchart;
3. how a nonterminating program behaves, according to the definition of M.

Note that M_{term} in Table 6.1.1 in line (i) refers to the evaluator for terms, and M_f in lines (iii)–(iv) refers to the evaluator for formulas (see Table 5.6.2).

Let us close this section by giving some examples of statements that do interesting things. Note that, just as a function body in Section 3.5 was a single expression (probably with many subexpressions), each of these programs is a single statement with many substatements. Note also that a statement is just a string, which we break into lines for readability.[4]

To multiply x and y, leaving the answer in z:

```
begin z←0;
      while y≠0 do begin
            z←z+x;
            y←y−1
            end
end
```

---

subproblems: compound statements correspond to sequential decomposition ("do this and then do that"); conditional statements correspond to decomposition by cases ("if the transaction is a withdrawal, do this; if it is a deposit, do that"); and while statements correspond to the decomposition "do one of many" ("for each input, do this" or "look at each element of the array until you find a such-and-so").

[3]It is pleasant to discover that the concept of an environment, that is, a function I: IVS→A (recall A is the set of values in the data type $\mathcal{A}$), which we introduced in Section 2.5, plays an important role in Chapters 3 and 5 and again here.

[4]In our definitions, we have continued the practice of underlining formal symbols; thus the entire statement ought to be underlined (just as we underlined expressions in Chapter 3). However, we bow to conventional usage in these examples, and underline only key words.

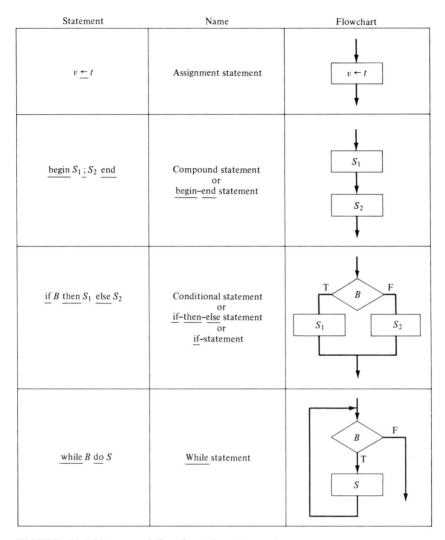

Statement	Name	Flowchart
$v \leftarrow t$	Assignment statement	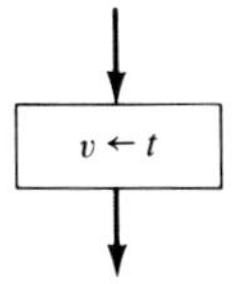
begin S_1 ; S_2 end	Compound statement or begin–end statement	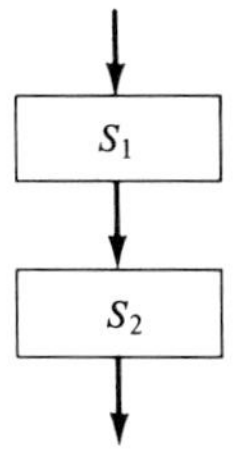
if B then S_1 else S_2	Conditional statement or if–then–else statement or if–statement	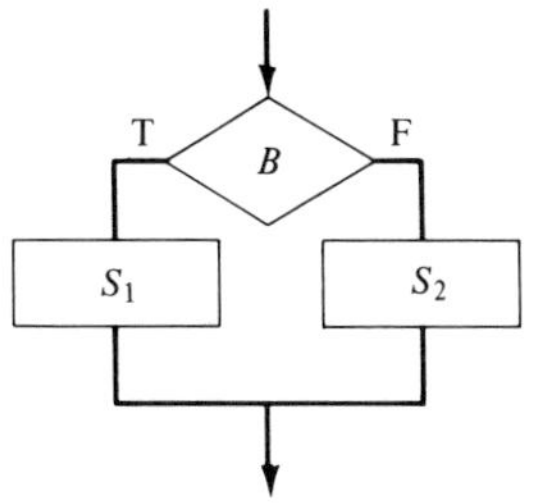
while B do S	While statement	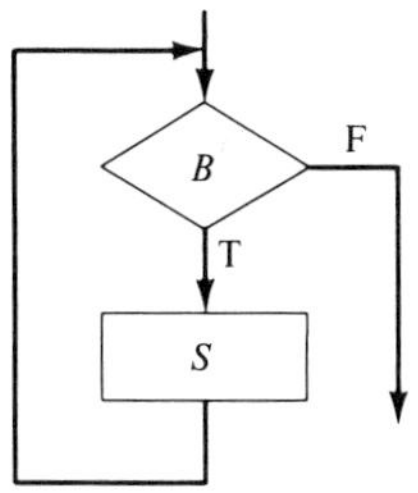

FIGURE 6.1.1 Names and flowcharts for statements.

TABLE 6.1.1 Definition of the evaluation function M

(i)	$M(I, v \leftarrow t) = I'$ where $I'(w) = \begin{cases} M_{\text{term}}(I,t) & \text{if } w = v \\ I(w) & \text{if } w \neq v \end{cases}$
(ii)	$M(I,$ begin S_1 ; S_2 end $) = M(M(I,S_1),S_2)$
(iii-t)	if $M_f(I,B) = \mathrm{T}$, then $M(I,$ if B then S_1 else $S_2) = M(I,S_1)$
(iii-f)	if $M_f(I,B) = \mathrm{F}$, then $M(I,$ if B then S_1 else $S_2) = M(I,S_2)$
(iv-f)	if $M_f(I,B) = \mathrm{F}$, then $M(I,$ while B do $S) = I$
(iv-t)	if $M_f(I,B) = \mathrm{T}$, then $M(I,$ while B do $S) = M(M(I,S),$ while B do $S)$

To calculate $z = \lfloor \sqrt{x} \rfloor$:

```
begin z←0;
      while (z+1)*(z+1) ≤ x do z←z+1
end
```

To reverse the stack x, leaving the answer in z:

```
begin z←empty;
      while ~ empty?(x) do
            if is0?(x) then begin z←push0(z);
                                  x←pop(x)
                            end
                       else begin z←push1(z);
                                  x←pop(x)
                            end
end
```

To calculate $z = \gcd(x,y)$:

```
begin while x≠y do
            if x<y then y←y−x
                   else x←x−y;
      z←x
end
```

EXERCISES 6.1

1. Draw flowcharts for the following statements:

(a)
```
begin z←0;
      while y≠0 do begin
            z←z+x;
            y←y−1
            end
end
```

(b)
```
begin z←0;
      while y≠0 do
            if even(y) then begin y←½y; x←z*x end
                       else begin y←y−1; z←z+x end
end
```

(c)
```
begin z←0;
      while y≠0 do
            begin
                  while even(y) do
                        begin y←½y; x←2*x end;
                  begin z←z+x; y←y−1 end
            end
end
```

2. Write a program that changes 0's to 1's and 1's to 0's on the stack x, leaving the answer in z.

*3. *Prove:* If $(I,S,I')\in M$ and $(I,S,I'')\in M$, then $I'=I''$ (i.e., statements are deterministic.)

*4. Extend the language of statements by adding a nondeterministic statement, such as $\underline{\text{choose}}\ S_1\ \underline{\text{or}}\ S_2\ \underline{\text{end}}$. Modify M appropriately.

6.2 THE LANGUAGE OF ASSERTIONS

In this section we shall introduce the language of *assertions* for expressing facts about the behavior of programs. We shall give an evaluation function for assertions that says what it means for an assertion to be true in a data type, but we shall see that it is impractical to work directly with the evaluation function. We shall therefore introduce a formal system, called *system H*, which will generate only true assertions.

Definition. If $\mathcal{Q}$ is a data type, a *partial correctness assertion* (or just *assertion*) over $\mathcal{Q}$ is a string of the form $P\underline{\{}S\underline{\}}Q$ where P and Q are first-order formulas over $\mathcal{Q}$ and S is a statement over $\mathcal{Q}$, such that no variable which appears in S is quantified in P or Q.[5]

Definition. The evaluation function M for assertions over $\mathcal{Q}$ is defined as follows:

$$M(P\underline{\{}S\underline{\}}Q)=\begin{cases}\mathrm{T} & \text{iff} \quad \text{for all environments } I \text{ and } I', \\ & \quad\quad \text{if } I'=M_s(I,S) \text{ and } M_f(I,P)=\mathrm{T}, \\ & \quad\quad \text{then } M_f(I',Q)=\mathrm{T} \\ \mathrm{F} & \text{otherwise}\end{cases}$$

We write $\mathcal{Q}\models P\underline{\{}S\underline{\}}Q$ if $M(P\underline{\{}S\underline{\}}Q)=\mathrm{T}$, and we say $P\underline{\{}S\underline{\}}Q$ is *true* in $\mathcal{Q}$.

Clearly, our first job is to figure out what all these different M's mean. The M in "$M(P\underline{\{}S\underline{\}}Q)=$" is the function we are trying to define. The M_f in "$M_f(I,P)$" or "$M_f(I',Q)$" is the evaluator for first-order formulas (Table 5.6.2), and M_s is the evaluation function for statements (Section 6.1).[6]

Now we can try to understand the definition. It says: If S starts at I and finishes at I', and P was true at I, then Q is true at I'. That is, if P was true before S started, then Q will be true after S finishes. There is only one small problem: What happens if S never terminates? In that case there is no I' such that $(I,S,I')\in M_s$, and the condition for $M(P\underline{\{}S\underline{\}}Q)=\mathrm{T}$ is true

[5]This restriction is entirely technical, and is put in to make the assignment rule in system H a little simpler.

[6]There, that wasn't so terrible, was it? For a somewhat harder example, look at lines 5 and 6 of Lemma 21 on page 206 of [Kleene 52] where the letter "A" appears in no fewer than five different typefaces (with five different meanings) within the space of two lines!

vacuously. So a better summary of "$\mathcal{A} \models P\{S\}Q$" is, "If P was true before, and S terminates, then Q is true afterward."[7]

Now we can formulate correctness assertions for some of our programs. In each case, we shall write down an assertion $P\{S\}Q$ that expresses the way we would *like* the program S to behave. To find out whether S is correct, we must evaluate $M(P\{S\}Q)$.

Multiplication program:

```
x = x₀ & y = y₀ {begin z ← 0;
                       while y ≠ 0 do begin
                             z ← z + x;
                             y ← y − 1
                             end
                 end} z = x₀ * y₀
```

Square root program:

```
x ≥ 0 {begin z ← 0;
             while (z + 1)*(z + 1) ≤ x do z ← z + 1
       end}
z² ≤ x & (z + 1)² > x
```

Greatest common divisor:

```
x = x₀ & y = y₀ & x > 0 & y > 0
{begin while x ≠ y do
             if x < y then y ← y − x
                      else y ← x − y;
       z ← x
 end}
(z | x₀) & (z | y₀) & (∀u)((u | x₀) & (u | y₀) ⊃ u ≤ z)
```

(In this example, $(x|y)$ means "x divides y evenly.")

Unfortunately, deciding whether $\mathcal{A} \models P\{S\}Q$ on the basis of the definition is an impossible task, since it means checking every pair of interpretations I, I' in $\mathcal{A}$.[8] Hence we introduce a formal system, called system H, with the property that if $P\{S\}Q \in H$, then $\mathcal{A} \models P\{S\}Q$.[9]

The objects of system H will be assertions over $\mathcal{A}$.[10]

[7]This is why $P\{S\}Q$ is called a *partial* correctness assertion. One could add a clause asserting termination, but that would make it much more difficult to use a system like system H. For discussions of total correctness systems, see [Dijkstra 76], [Manna and Waldinger 78], or [Manna 74, Section 3-3.3].

[8]Consequently, if we are to get anywhere, we need to do some mathematics. According to one seventh grader, "mathematics is finding the lazy man's way of doing something." Finding that way, of course, may be hard indeed!

TABLE 6.2.1 System H

$\dfrac{P \supset Q[^t_v]}{P\{v \leftarrow t\}Q}$	Assignment
$\dfrac{P\{S_1\}Q \quad Q\{S_2\}R}{P\{\ \underline{\text{begin}}\ S_1; S_2\ \underline{\text{end}}\ \}R}$	Compound
$\dfrac{P\&B\{S_1\}Q \quad P\&{\sim}B\{S_2\}Q}{P\{\ \underline{\text{if}}\ B\ \underline{\text{then}}\ S_1\ \underline{\text{else}}\ S_2\}Q}$	Conditional
$\dfrac{P \supset INV \quad INV\&B\{S\}INV \quad INV\&{\sim}B \supset Q}{P\{\ \underline{\text{while}}\ B\ \underline{\text{do}}\ S\}Q}$	While

Definition. System H is defined as follows:

(i) if $\mathcal{A} \models P \supset Q[^t_v]$, then $P\{v \leftarrow t\}Q \in H$
(ii) if $P\{S_1\}Q \in H$ and $Q\{S_2\}R \in H$, then $P\{\underline{\text{begin}}\ S_1; S_2\,\underline{\text{end}}\}R \in H$
(iii) if $P\&B\{S_1\}Q \in H$ and $P\&{\sim}B\{S_2\}Q \in H$, then $P\{\underline{\text{if}}\ B\ \underline{\text{then}}\ S_1\ \underline{\text{else}}\ S_2\}Q \in H$
(iv) if $\mathcal{A} \models P \supset INV$, $INV\&B\{S\}INV \in H$, and $\mathcal{A} \models INV\&{\sim}B \supset Q$, then $P\{\underline{\text{while}}\ B\ \underline{\text{do}}\ S\}Q \in H$
(v) nothing else

Here $Q[^t_v]$ means the formula Q with the term t substituted for each occurrence of the ivs v,[11] and in line (iv), INV stands for a formula.[12]

Using conventions similar to those used in the presentation of system G, we may write down system H in a convenient tabular format as in Table 6.2.1. Thus a derivation ("proof") in system H will be a tree (just as it was in system G) whose leaves are first-order formulas valid in $\mathcal{A}$. These formulas are called *verification conditions*. To show that $\mathcal{A} \models P\{S\}Q$, we use system H to generate verification conditions. We then must show, by whatever means available,[13] that the verification conditions are valid in the data type $\mathcal{A}$. If they are, we conclude that $\mathcal{A} \models P\{S\}Q$.

We may now state the plan of attack for the remainder of the chapter. In Section 3 we shall prove that if $P\{S\}Q \in H$, then $\mathcal{A} \models P\{S\}Q$. In the

---

[9]We name system H after C. A. R. Hoare, who invented the first system of this kind [Hoare 69]. Hoare's contribution was to put in axiomatic form some of the ideas of [Floyd 67].

[10]So, strictly speaking, we have a different system for each data type $\mathcal{A}$.

[11]Recall that "ivs" means "individual variable symbol." Substitution is discussed in Section 3.5 (especially footnote 27), and Section 5.3. See also the discussion of the assignment rule in Section 6.3.1.

[12]The reason for this choice will become clearer when we discuss the while-rule in Section 6.3.4.

[13]Usually by our "commonsense" knowledge about $\mathcal{A}$.

course of that section we shall also give an informal justification for each rule, as we did for system G. We shall also give examples of uses of the rules. In Section 4 we shall discuss how to apply system H. Unlike system G, system H requires some care in its application. In Section 5 we get to see (at long last!) some examples of complete proofs of assertions in system H. Finally, Sections 6 and 7 deal with how to use these ideas in writing programs.

EXERCISES 6.2

*1. If the language of statements is extended by adding a nondeterministic statement, as in Exercise 6.1.3, what does $M(P\{S\}Q)$ mean for nondeterministic programs?

2. If $\mathcal{Q} \models P\{S\}(x \neq x)$, what can we deduce about the behavior of S?

6.3 THE SOUNDNESS OF SYSTEM *H*

The goal of this section is to show that system H is a sound system for reasoning about programs, that is, if $P\{S\}Q \in H$, then $P\{S\}Q$ is a true assertion. Our proof will be similar to that of Theorem 5.9.1, which showed the soundness of system G. We shall show that the set of true assertions is closed under the rules of system H; therefore, by the fundamental theorem on induction (Theorem 2.3.1), the set H of assertions provable in system H is a subset of the set of true assertions.

Since we must show closure under each of the rules of system H, we shall discuss each rule in turn. For each rule, we shall present an informal justification or explanation, some examples, and a proof of soundness. The student who finds himself overwhelmed by the proofs (although they are not difficult) should nevertheless read the rest of the section carefully, for this is where the intuition about system H is to be developed.

6.3.1 The Assignment Rule

$$\frac{\mathcal{Q} \models P \supset Q\left[{}^{t}_{v}\right]}{P\{v \leftarrow t\}Q}$$

In order to determine the truth of $P\{v \leftarrow t\}Q$, it would be useful to have a formula Q' which could *predict* in a given environment I whether Q would be true in the environment I' after the assignment $v \leftarrow t$ was executed. More formally, if $(I, v \leftarrow t, I') \in M_s$, then $(\mathcal{Q}, I) \models Q'$ iff $(\mathcal{Q}, I') \models Q$. Now, I' is not much different from I (they differ only at the ivs v), so we might expect Q' to be not much different from Q. Indeed, so long as Q

does not use the variable v, we could let Q' be just Q itself. The only problem with this scheme arises when Q uses the variable v. Then Q' (evaluated at I) needs some way of predicting the value of v in I'. But $I'(v) = M_{\text{term}}(I,t)$, so we can predict (from I) the future value of v by simply evaluating the term t. So to compute the value of $M_f(I',Q)$, we can just compute $M_f(I,Q)$, except that every time we need to evaluate the variable v, we evaluate t instead. But that is what we do when we evaluate $M_f(I,Q[^t_v])$.

This will become clearer when we look at some examples.

EXAMPLE 1. Let $v \leftarrow t$ be $x \leftarrow x+1$, and let Q be the formula $x=a$. Q will be true after $x \leftarrow x+1$ iff $x+1=a$ before the assignment. $Q[^{x+1}_{x}]$ is $x+1=a$.

EXAMPLE 2. Let $v \leftarrow t$ be $y \leftarrow y+d$, and let Q be $y=(x+2)^2$. Then Q is true after the assignment iff $y+d=(x+2)^2$ is true before the assignment.

Thus $Q[^t_v]$ is true before $v \leftarrow t$ iff Q is true after $v \leftarrow t$. So if $P \supset Q[^t_v]$ is valid in $\mathscr{Q}$, and P is true before $v \leftarrow t$, then $Q[^t_v]$ is true also, and therefore Q is true afterward. This completes the informal justification of the rule.

EXAMPLE 3. Let

P	be	$x+1=a$
Q	be	$x=a$
$v \leftarrow t$	be	$x \leftarrow x+1$

The corresponding instance of the rule is

$$\frac{(x+1=a) \supset (x+1=a)}{x+1=a\{x \leftarrow x+1\}x=a}$$

Here P is just $Q[^t_v]$. The verification condition is valid in any data type.

EXAMPLE 4. To show true$\{x \leftarrow 3\}x=3$ (here "true" is a formula which is always true):

$$\frac{\text{true} \supset 3=3}{\text{true}\{x \leftarrow 3\}x=3}$$

Q is $x=3$, so $Q[^t_v]$ is $3=3$. Again, the verification condition is valid in any data type.

EXAMPLE 5

$$\frac{(x>0)\,\&\,(y \leqslant x) \supset (y \dot{-} 1 < x)}{(x>0)\,\&\,(y \leqslant x)\{y \leftarrow y \dot{-} 1\}y<x}$$

Here the verification condition is valid in the data type $\mathcal{N}$ of the nonnegative integers (Section 3.1). If the $x>0$ were omitted, then the verification condition would not be valid (it is false at $y=x=0$; there $y \dot{-} 1=0$), and the partial correctness assertion would likewise be false.

We now proceed to a more rigorous proof of the soundness of the assignment rule.

Theorem 6.3.1. *Let v be a variable, t a term, and I an environment. Let I' be defined by*

$$I'(w)=\begin{cases} I(w) & w\neq v \\ M_{\text{term}}(I,t) & w=v \end{cases}$$

Then

(i) *If u is any term, then $M_{\text{term}}(I',u)=M_{\text{term}}(I,u[{}^{t}_{v}])$.*
(ii) *If G is an atomic formula, then $M_{\text{f}}(I',u)=M_{\text{f}}(I,G[{}^{t}_{v}])$.*
(iii) *Let G be any formula in which neither v nor any variable appearing in t appears quantified. Then $M_{\text{f}}(I',G)=M_{\text{f}}(I,G[{}^{t}_{v}])$.*
(iv) *Let P and Q be formulas restricted as in (iii). If $\mathcal{A}\vDash P\supset Q$, then $\mathcal{A}\vDash P\{v\leftarrow t\}Q$.*

PROOF. (i) By induction on the construction of a term. We write M for M_{term}. If c is a constant symbol, then $c[{}^{t}_{v}]$ is just c, and then $M(I',c)=c^{\mathcal{A}}=M(I,c)=M(I,c[{}^{t}_{v}])$. If w is an individual variable symbol other than v, then $w[{}^{t}_{v}]=w$ and $M(I',w)=I'(w)=I(w)=M(I,w)=M(I,w[{}^{t}_{v}])$. To finish the base case, observe that $v[{}^{t}_{v}]$ is t, so $M(I',v)=I'(v)=M(I,t)=M(I,v[{}^{t}_{v}])$. For the induction step, let $u=f\underline{(}u_1\underline{,}\ldots\underline{,}u_n\underline{)}$ where f is an n-place function symbol, and $u_1,\ldots,u_n$ are terms. Then

$$\begin{aligned} M(I',f\underline{(}u_1\underline{,}\ldots\underline{,}u_n\underline{)}) &= f^{\mathcal{A}}(M(I',u_1),\ldots,M(I',u_n)) && \text{(defn. of } M) \\ &= f^{\mathcal{A}}(M(I,u_1[{}^{t}_{v}]),\ldots,M(I,u_n[{}^{t}_{v}])) && \text{(by IH)} \\ &= M(I,f\underline{(}u_1[{}^{t}_{v}]\underline{,}\ldots\underline{,}u_n[{}^{t}_{v}]\underline{)}) && \text{(defn. of } M) \\ &= M(I,u[{}^{t}_{v}]) && \text{(defn. of substitution)} \end{aligned}$$

This completes the induction for (i).

(ii) If G is an atomic formula, then $G=p\underline{(}u_1\underline{,}\ldots\underline{,}u_n\underline{)}$ for some n-place predicate symbol p and n terms $u_1,\ldots,u_n$. Then

$$\begin{aligned} M_{\text{f}}(I',G) &= p^{\mathcal{A}}(M(I',u_1),\ldots,M(I',u_n)) \\ &= p^{\mathcal{A}}(M(I,u_1[{}^{t}_{v}]),\ldots,M(I,u_n[{}^{t}_{v}])) && \text{(by (i))} \\ &= M_{\text{f}}(I,p\underline{(}u_1[{}^{t}_{v}]\underline{,}\ldots\underline{,}u_n[{}^{t}_{v}]\underline{)}) && \text{(defn. of } M_{\text{f}}) \\ &= M_{\text{f}}(I,G[{}^{t}_{v}]) && \text{(defn. of substitution)} \end{aligned}$$

(iii) Left as an exercise.

(iv) Assume $\mathcal{A}\vDash P\supset Q[{}^{t}_{v}]$ and $M_{\text{f}}(I,P)=\text{TRUE}$. We must show that if $J=M_{\text{s}}(I,v\leftarrow t)$, then $M_{\text{f}}(J,Q)=\text{TRUE}$. If $J=M_{\text{s}}(I,v\leftarrow t)$, then (by the definition of M_s) J must be the I' in the statement of the theorem. If $M_f(I,P)=\text{TRUE}$, and $P \supset Q[^t_v]$ is valid in $\mathfrak{A}$, then $M_f(I,Q[^t_v])=\text{TRUE}$. Therefore, by (iii), $M_f(I',Q)=\text{TRUE}$. □

6.3.2 The Compound Rule

$$\frac{P\{S_1\}Q \quad Q\{S_2\}R}{P\{\underline{\text{begin}}\ S_1;\ S_2\ \underline{\text{end}}\}R}$$

After the assignment rule (which we admit is at least a trifle obscure), the compound rule should come as some relief. We assume that over our data type $\mathfrak{A}$ we have a statement S_1 such that if when it starts the formula P is true, then after it finishes the formula Q is true (that is, $\mathfrak{A} \vDash P\{S_1\}Q$); and a second statement S_2, which, started in an environment where Q is true, finishes (if it halts) in an environment where R is true (that is, $\mathfrak{A} \vDash Q\{S_2\}R$).

We wish to conclude that $\mathfrak{A} \vDash P\{\underline{\text{begin}}\ S_1; S_2\ \underline{\text{end}}\}R$. To do this, imagine starting the execution of $\underline{\text{begin}}\ S_1; S_2\ \underline{\text{end}}$ in an environment in which P is true. We first execute S_1, resulting in an environment where Q is true (since $\mathfrak{A} \vDash P\{S_1\}Q$). We then execute S_2. Since $Q\{S_2\}R$ is true in $\mathfrak{A}$, when (and if) S_2 terminates, the resulting environment will make R true. So we have deduced that after the execution of $\underline{\text{begin}}\ S_1; S_2\ \underline{\text{end}}$ we will be in an environment making R true. So $P\{\underline{\text{begin}}\ S_1; S_2\ \underline{\text{end}}\}R$ is true in $\mathfrak{A}$.

EXAMPLE 6

$$\frac{\begin{array}{c}(x>0)\&(x+y=a)\{x\leftarrow x-1\}(x+(y+1)=a)\\ (x+(y+1)=a)\{y\leftarrow y+1\}(x+y)=a\end{array}}{(x>0)\&(x+y=a)\{\ \underline{\text{begin}}\ x\leftarrow x-1; y\leftarrow y+1\ \ \underline{\text{end}}\ \}(x+y=a)}$$

Here we have listed the assumptions of the rule vertically since they would not fit on the page otherwise. The intermediate formula Q is $x+(y+1)=a$, which is $R[^{y+1}_{y}]$. We can complete the derivation of this assertion in system H by applying the assignment rule to each of the hypotheses:

$$\frac{\dfrac{(x>0)\&(x+y=a)\supset((x-1)+(y+1)=a)}{(x>0)\&(x+y=a)\{x\leftarrow x-1\}(x+(y+1)=a)} \quad \dfrac{x+(y+1)=a\supset x+(y+1)=a}{x+(y+1)=a\{y\leftarrow y+1\}(x+y=a)}}{(x>0)\&(x+y=a)\{\ \underline{\text{begin}}\ x\leftarrow x-1; y\leftarrow y+1\ \underline{\text{end}}\ \}(x+y=a)}$$

For more on how to choose the intermediate formula Q, see Section 4. For the moment, we can proceed with the formal proof:

Theorem 6.3.2. *If $\mathfrak{A} \vDash P\{S_1\}Q$ and $\mathfrak{A} \vDash Q\{S_2\}R$, then*

$$\mathfrak{A} \vDash P\{\ \underline{\text{begin}}\ \ S_1; S_2\ \underline{\text{end}}\ \}R$$

PROOF. Let I be an environment such that $M_f(I,P)=\text{TRUE}$. We need to show that if $I''=M_s(I,\underline{\text{begin}}\,S_1;S_2\,\underline{\text{end}})$, then $M_f(I'',R)=\text{TRUE}$. By the definition of M_s (Table 6.1.1), $I''=M_s(M_s(I,S_1),S_2)$. Let $I'=M_s(I,S_1)$. Since $\mathcal{Q}\vDash P\{S_1\}Q$, $M_f(I',Q)=\text{TRUE}$. But then since $I''=M_s(I',S_2)$ and $\mathcal{Q}\vDash Q\{S_2\}R$, $M_f(I'',R)=\text{TRUE}$. □

6.3.3 The Conditional Rule

$$\frac{P\&B\{S_1\}Q \quad P\&\sim B\{S_2\}Q}{P\{\underline{\text{if}}\ B\ \underline{\text{then}}\ S_1\ \underline{\text{else}}\ S_2\}Q}$$

Again, the conditional rule is moderately simple. Imagine starting the program $\underline{\text{if}}\ B\ \underline{\text{then}}\ S_1\ \underline{\text{else}}\ S_2$ in an environment where P is true. If B is true, then S_1 is executed; we know P and B are true on entry to S_1, so (since $\mathcal{Q}\vDash P\&B\{S_1\}Q$) Q will be true on exit. On the other hand, if B is false, then S_2 is executed, we know $P\&\sim B$ is true on entry to S_2, so Q will be true on exit. Either way, we are guaranteed that Q is true on exit from the conditional statement.

EXAMPLE 7. In the data type of nonnegative integers $\mathcal{N}$:

$$\frac{\dfrac{(y\leqslant x)\&(x=0)\supset(y<x+1)}{(y\leqslant x)\&(x=0)\{x\leftarrow x+1\}(y<x)} \quad \dfrac{(y\leqslant x)\&(x\neq 0)\supset((y\dot{-}1)<x)}{(y\leqslant x)\&(x\neq 0)\{y\leftarrow y\dot{-}1\}(y<x)}}{(y\leqslant x)\{\underline{\text{if}}\ x=0\ \underline{\text{then}}\ x\leftarrow x+1\ \underline{\text{else}}\ y\leftarrow y\dot{-}1\}(y<x)}$$

EXAMPLE 8

$$\frac{\begin{array}{c}\text{true}\&(x\neq probe)\{found\leftarrow 0\}(found=1\Leftrightarrow x=probe)\\ \text{true}\&\sim(x\neq probe)\{found\leftarrow 1\}(found=1\Leftrightarrow x=probe)\end{array}}{\text{true}\{\underline{\text{if}}\ x\neq probe\ \underline{\text{then}}\ found\leftarrow 0\ \underline{\text{else}}\ found\leftarrow 1\}(found=1\Leftrightarrow x=probe)}$$

We have written $G\Leftrightarrow H$ for $(G\supset H)\&(H\supset G)$. Note that in the second assumption, the precondition is true $\&\sim(x\neq probe)$ rather than true$\&(x=probe)$; the rule says the precondition is $P\&\sim B$, even if $\sim B$ is logically equivalent to some simpler expression!

Theorem 6.3.3. *If* $\mathcal{Q}\vDash P\&B\{S_1\}Q$ *and* $\mathcal{Q}\vDash P\&\sim B\{S_2\}Q$, *then*

$$\mathcal{Q}\vDash P\{\underline{\text{if}}\ B\ \underline{\text{then}}\ S_1\ \underline{\text{else}}\ S_2\}Q$$

PROOF. Choose I such that $M_f(I,P)=\text{TRUE}$. We must show that if $I'=M_s(I,\underline{\text{if}}\,B\,\underline{\text{then}}\,S_1\,\underline{\text{else}}\,S_2)$, then $M_f(I',Q)=\text{TRUE}$. There are two cases:

Case I: $M_f(I,B)=\text{TRUE}$. Then $I'=M_s(I,S_1)$, and since $\mathcal{Q}\vDash P\&B\{S_1\}Q$, $M_f(I',Q)=\text{TRUE}$.

Case II: $M_f(I,B)=\text{FALSE}$. Then $I'=M_s(I,S_2)$, and since $\mathfrak{A}\vDash P\&\sim B\{S_2\}Q$, $M_f(I',Q)=\text{TRUE}$. □

6.3.4 The While Rule

$$\frac{\mathfrak{A}\vDash P\supset INV \quad INV\&B\{S\}INV \quad \mathfrak{A}\vDash INV\&\sim B\supset Q}{P\{\underline{\text{while}}\ B\ \underline{\text{do}}\ S\}Q}$$

The crucial thing about understanding the while-rule is the formula *INV*. This formula is called the *invariant* of the loop, and the hypotheses of the rule imply that the invariant is true every time control reaches the top of the loop (by which we mean the loop test *B*). Let us see why this is so. The argument goes by induction:

Imagine starting the program while *B* do *S* in an environment where *P* is true (Figure 6.3.1). Since $P\supset INV$ is valid in $\mathfrak{A}$, *INV* is true the first time control reaches the loop test *B*.

Now imagine that *INV* is true the *k*th time control reaches the loop test. If the loop test *B* is false, then we exit, with *INV* still true. If the loop test *B* is true, then we execute the loop body *S* in the current environment, in which $INV\&B$ is true. Since $\mathfrak{A}\vDash INV\&B\{S\}INV$, we reach the top of the loop for the $(k+1)$th time in an environment where *INV* is true once more.

So, if we start the loop in an environment where *P* is true, then every time control reaches the top of the loop, *INV* is true. In particular, *INV* is true the last time we reach the top of the loop, that is, the time when *B* is false. So we exit the loop with $INV\&\sim B$ true. Since $INV\&\sim B\supset Q$ is

FIGURE 6.3.1 Annotated flowchart for a while loop.

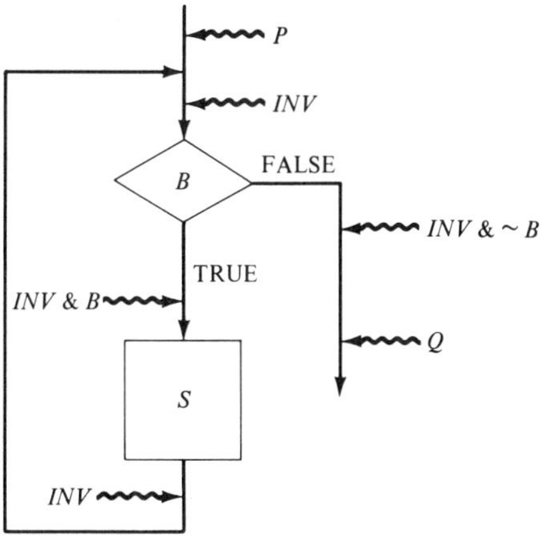

valid in $\mathcal{Q}$, Q is true on exit. This completes the argument for the while-rule.

EXAMPLE 9

$$\frac{\begin{array}{c}(x=x_0)\&(y=y_0)\&(z=0)\supset(x*y+z=x_0*y_0)\\ (x*y+z=x_0*y_0)\&(y\neq 0)\{\underline{\text{begin}}\ z\leftarrow z+x;y\leftarrow y-1\ \underline{\text{end}}\}(x*y+z=x_0*y_0)\\ (x*y+z=x_0*y_0)\&\sim(y\neq 0)\supset(z=x_0*y_0)\end{array}}{\begin{array}{l}(x=x_0)\&(y=y_0)\&(z=0)\\ \quad\{\underline{\text{while}}\ y\neq 0\ \underline{\text{do}}\\ \qquad\underline{\text{begin}}\ z\leftarrow z+x;y\leftarrow y-1\ \underline{\text{end}}\}(z=x_0*y_0)\end{array}}$$

Here the loop invariant is $(x*y+z=x_0*y_0)$; again we have listed the antecedents of the rule vertically. Now, it is by no means obvious where the invariant came from; indeed, an invariant represents deep knowledge about how a program works, as we shall see in later sections. In the meantime, we proceed with the soundness proof.

Theorem 6.3.4. *If* $\mathcal{Q}\vDash P\supset INV$, $\mathcal{Q}\vDash INV\&B\{S\}INV$, *and* $\mathcal{Q}\vDash INV\&\sim B\supset Q$, *then* $\mathcal{Q}\vDash P\{\underline{\text{while}}\,B\,\underline{\text{do}}\,S\}Q$.

PROOF. The proof proceeds by induction on the construction of the meaning function M_s as a relation in stages $M_s^{(k)}$, as in Section 2.2. The induction hypothesis is:

IH_k: For all I,I'', if $M_f(I,INV)=\text{TRUE}$ and $(I,\ \underline{\text{while}}\ B\ \underline{\text{do}}\ S,I'')\in M_s^{(k)}$, then $M_f(I'',Q)=\text{TRUE}$.

Base Step: $(k=0)$: The only way $(I,\underline{\text{while}}\,B\,\underline{\text{do}}\,S,I'')$ can belong to $M_s^{(0)}$ is by rule (iv-f) of Table 6.1.1; in that case $M_f(I,B)=\text{FALSE}$ and $I''=I$. Hence $M_f(I'',INV\&\sim B)=M_f(I,INV\&\sim B)=\text{TRUE}$. Since $INV\&\sim B\supset Q$ is valid in $\mathcal{Q}$, $M_f(I'',Q)=\text{TRUE}$ also.

Induction Step: Assume IH_k, and $(I,\underline{\text{while}}\,B\,\underline{\text{do}}\,S,I'')\in M_s^{(k+1)}$. If $(I,\underline{\text{while}}\,B\,\underline{\text{do}}\,S,I'')\in M_s^{(k)}$, we are done, by IH_k; otherwise, according to Table 6.1.1, we must have $M_f(I,B)=\text{TRUE}$ and there must be an I' such that $(I,S,I')\in M_s^{(k)}$ and $(I',\underline{\text{while}}\,B\,\underline{\text{do}}\,S,I'')\in M_s^{(k)}$. Since $\mathcal{Q}\vDash INV\&B\{S\}INV$, $M_f(I',INV)=\text{TRUE}$. Hence we can apply IH_k to I', deducing that $M_f(I'',Q)=\text{TRUE}$.

From the induction, we conclude $\mathcal{Q}\vDash INV\{\underline{\text{while}}\,B\,\underline{\text{do}}\,S\}Q$; since $\mathcal{Q}\vDash P\supset INV$, we deduce that $\mathcal{Q}\vDash P\{\underline{\text{while}}\,B\,\underline{\text{do}}\,S\}Q$. □

Theorem 6.3.5. *If* $P\{S\}Q\in H$, *then* $\mathcal{Q}\vDash P\{S\}Q$.

PROOF. Theorems 6.3.1–6.3.4 show that the set of assertions valid in $\mathcal{Q}$ is closed under the rules of system H; by the fundamental theorem on induction (Theorem 2.3.1) H is then a subset of the set of assertions valid in $\mathcal{Q}$. □

Unfortunately, the converse of Theorem 6.3.5 requires some additional assumptions about the data type $\mathcal{Q}$ (see [Wand 78], [Cook 78]). Nor does the analogue of Theorem 5.9.2 hold: One can choose an intermediate assertion or loop invariant badly and wind up in a blind alley trying to verify a provable assertion. It is this problem we shall attack in the next section.

EXERCISES 6.3

1. Finish the proof of Theorem 6.3.1(iii). (*Hint:* Watch out for the restriction on quantified variables.)

6.4 USING SYSTEM H

Given an assertion $P\{S\}Q$, how do we go about determining whether it is in H? Just as we did in system G, we construct a derivation tree whose root is $P\{S\}Q$ and whose leaves are first-order formulas valid in $\mathcal{Q}$. How do we go about constructing the tree? Given an assertion, we attempt to locate a rule by which the assertion could have been introduced into H. Now we have some good news and some bad news: The good news is that it is easy to find the rule that was used, because there is exactly one rule per statement type. The bad news is that this is not always enough to build the tree. Take, for an example, an assertion of the form $x=2$ $\{\underline{\text{begin}}\, S_1; S_2\, \underline{\text{end}}\}\, y=4$. It can get into H only by an instance of the compound statement rule

$$\frac{P\{S_1\}Q \quad Q\{S_2\}R}{P\{\ \underline{\text{begin}}\ S_1; S_2\ \underline{\text{end}}\ \}R}$$

where P is "$x=2$" and R is "$y=4$." But what should we choose for Q? The rule gives us no clue.

In system H, unlike system G, we may not blithely choose an arbitrary Q and still be assured of success. For example we could proceed as follows:

$$\frac{x=2\{y\leftarrow 4\}y=3 \quad y=3\{x\leftarrow 7\}y=4}{x=2\{\ \underline{\text{begin}}\ y\leftarrow 4; x\leftarrow 7\ \underline{\text{end}}\ \}y=4}$$

but both $x=2\{y\leftarrow 4\}y=3$ and $y=3\{x\leftarrow 7\}y=4$ are false in $\mathfrak{N}$, so they cannot be in H. So we have reached a dead end. But if we choose "$y=4$" for Q, we win:

$$\frac{\dfrac{(x=2)\supset(4=4)}{x=2\{y\leftarrow 4\}y=4} \quad \dfrac{(y=4)\supset(y=4)}{y=4\{x\leftarrow 7\}y=4}}{x=2\{\ \underline{\text{begin}}\ y\leftarrow 4; x\leftarrow 7\ \underline{\text{end}}\ \}y=4}$$

If we look at System H, we discover that two of the four rules require us to "fill in the blanks" in this manner: the compound rule

$$\frac{P\{S_1\}Q \quad Q\{S_2\}R}{P\{\ \underline{\text{begin}}\ S_1; S_2\ \underline{\text{end}}\ \}R}$$

and the while-rule

$$\frac{P\supset INV \quad INV\&B\{S\}INV \quad INV\&\sim B\supset Q}{P\{\ \underline{\text{while}}\ B\ \underline{\text{do}}\ S\}Q}$$

In general, it is rather difficult to come up with a loop invariant. Loop invariants represent some deep understanding of the program, knowledge that it is unreasonable to provide mechanically. Hence we shall usually supply the invariant with the program (as a mandatory comment) and write

$$\underline{\text{while}}\ B\ \underline{\text{inv}}\ I\ \underline{\text{do}}\ S$$

with the rule

$$\frac{P\supset I \quad I\&B\{S\}I \quad I\&B\supset Q}{P\{\ \underline{\text{while}}\ B\ \underline{\text{inv}}\ I\ \underline{\text{do}}\ S\}Q}$$

This is not so unreasonable, since, as we shall see, a good programmer generates her invariants as she goes along. As Wirth remarks: "The lesson every programmer should learn is that the *explicit indication of the relevant invariant for each repetition represents the most valuable element in every program documentation*" ([Wirth 73, p. 23] emphasis in original).

In this section we shall present some techniques for finding the intermediate formula Q. These techniques are "guaranteed" in the sense that if $P\{\underline{\text{begin}}\,S_1; S_2\,\underline{\text{end}}\}R$ is provable in system H, then it is provable with the Q supplied by the technique.

Technique 6.4.1. *To prove* $P\{\underline{\text{begin}}\,v\leftarrow t; S_2\,\underline{\text{end}}\}R$ *where* P *and* t *do not contain the variable* v, *choose* Q *to be* $P\&(v=t)$.

By this technique, the proof tree looks like this:

$$\frac{\dfrac{P\supset(P\&v=t)\left[\begin{smallmatrix}t\\v\end{smallmatrix}\right]}{P\{v\leftarrow t\}P\&v=t} \quad P\&v=t\{S_2\}R}{P\{\ \underline{\text{begin}}\ v\leftarrow t; S_2\ \underline{\text{end}}\ \}R}$$

Since P and t do not contain the variable v, we have $(P\&v=t)[^t_v]=(P[^t_v]\&(v=t)[^t_v])=(P\&(t=t))$, so the verification condition is $P\supset(P\&(t=t))$, which is valid, and is therefore valid in any data type $\mathcal{Q}$. Note that if P or t contains v, then this technique fails miserably. Take

$$x=2\{\ \underline{\text{begin}}\ x\leftarrow 3; S_2\ \underline{\text{end}}\ \}R$$

If we try to apply Technique 6.4.1, we get

$$\frac{x=2\{x\leftarrow 3\}x=2\&x=3 \quad x=2\&x=3\{S_2\}R}{x=2\{\ \underline{\text{begin}}\ x\leftarrow 3;S_2\ \underline{\text{end}}\ \}R}$$

Here we have come up with the subgoal of showing that if we start with $x=2$ and then execute the statement $x\leftarrow 3$, then afterwards x equals both 2 and 3. Since this is palpably false, it is not in H, and hence it is not a good choice for a subgoal!

Technique 6.4.1 shows our general strategy for choosing Q's: by a proper choice of Q, we arrive at a verification condition which is guaranteed to be a valid formula. Hence we can concentrate our efforts on the other subgoal(s).

Technique 6.4.2. *To prove* $P\{\underline{\text{begin}}\,S_1;v\leftarrow t\,\underline{\text{end}}\}R$, *choose* Q *to be* $R[{}^t_v]$.

The proof tree is

$$\frac{P\{S_1\}R[{}^t_v] \quad \dfrac{R[{}^t_v]\supset R[{}^t_v]}{R[{}^t_v]\{v\leftarrow t\}R}}{P\{\ \underline{\text{begin}}\ S_1;v\leftarrow t\ \underline{\text{end}}\ \}R}$$

Here we get one valid verification condition $R[{}^t_v]\supset R[{}^t_v]$ and one subgoal $P\{S_1\}R[{}^t_v]$.

Technique 6.4.3. *To prove* $P\{\underline{\text{begin}}\,S_1;\underline{\text{while}}\,B\,\underline{\text{inv}}\,I\,\underline{\text{do}}\,S_2\underline{\text{end}}\}R$, *choose* Q *to be* I.

The proof tree is

$$\frac{P\{S_1\}I \quad \dfrac{I\supset I \quad I\&B\{S_2\}I \quad I\&\sim B\supset R}{I\{\ \underline{\text{while}}\ B\ \underline{\text{inv}}\ I\ \underline{\text{do}}\ S_2\}R}}{P\{\ \underline{\text{begin}}\ S_1;\ \underline{\text{while}}\ B\ \underline{\text{inv}}\ I\ \underline{\text{do}}\ S_2\ \underline{\text{end}}\ \}R}$$

Here we arrive at one guaranteed valid verification condition, $I\supset I$, one "real" verification condition, $I\&\sim B\supset R$, and two subgoals, $P\{S_1\}I$ and $I\&B\{S_2\}I$.

A similar analysis can be performed for any compound statement; these are discussed in the exercises.

Another way of looking at these techniques is as "derived rules," which are abbreviations for the pieces of proof tree shown here. This point of view has the advantage that it avoids generating verification conditions which are valid and therefore uninteresting. Table 6.4.1 shows these three techniques as derived rules of system H.

TABLE 6.4.1 Some derived rules for compound statements

$$\frac{P \& v=t\{S_2\}R}{P\{\ \underline{\text{begin}}\ v\leftarrow t;\, S_2\ \underline{\text{end}}\ \}R}\ \text{init(if } v \text{ does not occur in } P \text{ or } t\text{)}$$

$$\frac{P\{S_1\}R\left[{}^{t}_{v}\right]}{P\{\ \underline{\text{begin}}\ S_1;\, v\leftarrow t\ \underline{\text{end}}\ \}R}\ ;\leftarrow$$

$$\frac{P\{S_1\}I \quad I\&B\{S_2\}I \quad I\&\sim B\supset R}{P\{\ \underline{\text{begin}}\ S_1;\ \underline{\text{while}}\ B\ \underline{\text{inv}}\ I\ \underline{\text{do}}\ S_2\ \underline{\text{end}}\ \}R}\ ;\text{while}$$

Now that the derived rules are available, there may be more than one rule applicable to an assertion. For example, take an initialization followed by a loop:

$$P\{\ \underline{\text{begin}}\ v\leftarrow t;\ \underline{\text{while}}\ B\ \underline{\text{inv}}\ I\ \underline{\text{do}}\ S\ \underline{\text{end}}\ \}R$$

where v does not occur in P or t. Then the compound statement rule and the first and third derived rules in Table 6.4.1 are all possible rules to use. Using the first derived rule, we get:

$$\cfrac{\cfrac{P\&(v=t)\supset I \quad I\&B\{S\}I \quad I\&\sim B\supset R}{P\&v=t\{\ \underline{\text{while}}\ B\ \underline{\text{inv}}\ I\ \underline{\text{do}}\ S\}R}}{P\{\ \underline{\text{begin}}\ v\leftarrow t;\ \underline{\text{while}}\ B\ \underline{\text{inv}}\ I\ \underline{\text{do}}\ S\ \underline{\text{end}}\ \}R}$$

Using the third derived rule gives a slightly different set of verification conditions.

EXERCISES 6.4

Demonstrate the correctness of the following derived rules of inference:

***1.** $$\frac{P\supset I \quad I\&B\{S_1\}I \quad I\&\sim B\{S_2\}Q}{P\{\underline{\text{begin}}\,\underline{\text{while}}\,B\,\underline{\text{inv}}\,I\,\underline{\text{do}}\,S_1;\,S_2\,\underline{\text{end}}\}\,Q}\ \text{while;}$$

Show that the following rules are admissible, that is, if all the hypotheses are provable in H, so is the conclusion.

2. $$\frac{P\{\ \underline{\text{begin}}\ \ \underline{\text{begin}}\ S_1;\, S_2\ \underline{\text{end}}\ ;\, S_3\ \underline{\text{end}}\ \}}{P\{\underline{\text{begin}}\,S_1;\,\underline{\text{begin}}\,S_2;\,S_3\,\underline{\text{end}}\,\underline{\text{end}}\}R}\ ;;$$

***3.** $$\frac{P\{S\}Q \quad Q\supset R}{P\{S\}R}\ \text{Conseq}_R$$

***4.** $$\frac{P\supset Q \quad Q\{S\}R}{P\{S\}R}\ \text{Conseq}_L$$

***5.** $$\frac{P\{S_1\}(B \supset Q_2)\&(\sim B \supset Q_3) \quad Q_2\{S_2\}R \quad Q_3\{S_3\}R}{P\{\underline{\text{begin}}\, S_1;\, \underline{\text{if}}\, B\, \underline{\text{then}}\, S_2\, \underline{\text{else}}\, S_3\, \underline{\text{end}}\}R}\ ;\text{if}$$

(*Hint:* Use Conseq_L.)

***6.** *Prove:* If no variable of S appears in P, then $P\{S\}P \in H$.

7. Give a *different* derived rule for an initialized while, as suggested in the text. What is the difference between the two?

***8.** For each of the derived rules in Table 6.4.1, show that if there is *any* proof of the conclusion in H, then there is a proof using the derived rule.

6.5 EXAMPLES IN SYSTEM H

Our first example is a simple addition program:

$$\begin{array}{l}(x=x_0)\&(y=y_0)\\ \{\underline{\text{while}}\, x \neq 0\, \underline{\text{inv}}\, x+y=x_0+y_0\\ \qquad\qquad \underline{\text{do}}\, \underline{\text{begin}}\, x \leftarrow x-1;\\ \qquad\qquad\qquad\quad y \leftarrow y+1\\ \qquad\qquad\quad \underline{\text{end}}\}\\ y=x_0+y_0\end{array}$$

The proof tree is given in Figure 6.5.1. The verification conditions are:

$$\begin{array}{ll}\text{VC\#1:} & (x=x_0)\&(y=y_0)\supset(x+y=x_0+y_0)\\ \text{VC\#2:} & (x+y=x_0+y_0)\&(x\neq 0)\supset((x-1)+(y+1)=(x_0+y_0))\\ \text{VC\#3:} & (x+y=x_0+y_0)\&\sim(x\neq 0)\supset(y=x_0+y_0)\end{array}$$

All three VC's are true in $\mathfrak{N}$, so the assertion is true in $\mathfrak{N}$: the program works![14] Unfortunately, it is clear that if we persist in presenting system H proofs in the format of Figure 6.5.1, we shall rapidly exceed the capacity of most humans to organize information on the printed page.[15] Hence we adopt some useful conventions:

1. We allow abbreviations for formulas and statements.
2. We carry substitutions around, rather than making them immediately.
3. We return to trees, rather than using the unwieldly "horizontal line" format.

Figure 6.5.2 shows the result of rewriting Figure 6.5.1 in the new format. We can now look at the proof.

ENTRY$\{$*PROG*$\}$*EXIT* is the abbreviated form of the assertion we are trying to prove. We applied the while-rule to this assertion, yielding two

[14] Thus we have succeeded in formalizing Exercise 2.1.7.

[15] For example, in Figure 6.5.1, what rule does the middle horizontal line represent?

$$\cfrac{\cfrac{\cfrac{(x+y=x_0+y_0)\,\&\,(x\neq 0)\supset((x-1)+(y+1)=(x_0+y_0))}{(x+y=x_0+y_0)\,\&\,(x\neq 0)\{x\leftarrow x-1\}(x+(y+1)=x_0+y_0)}}{(x=x_0)\,\&\,(y=y_0)\supset(x+y=x_0+y_0)\quad(x+y=x_0+y_0)\,\&\,(x\neq 0)\{\ \underline{\text{begin}}\ x\leftarrow x-1;y\leftarrow y+1\ \underline{\text{end}}\ \}(x+y=x_0+y_0)\quad(x+y=x_0+y_0)\,\&\sim(x\neq 0)\supset(y=x_0+y_0)}}{(x=x_0)\,\&\,(y=y_0)\{\ \underline{\text{while}}\ x\neq 0\ \underline{\text{inv}}\ x+y=x_0+y_0\ \underline{\text{do}}\ \ \underline{\text{begin}}\ x\leftarrow x-1;y\leftarrow y+1\ \underline{\text{end}}\ \}\,y=x_0+y_0}$$

FIGURE 6.5.1

Abbreviations

ENTRY	is	$(x=x_0)\&(y=y_0)$
EXIT	is	$y=x_0+y_0$
INV	is	$x+y=x_0+y_0$
PROG	is	while $x\neq 0$ inv *INV* do begin $x\leftarrow x-1; y\leftarrow y+1$ end

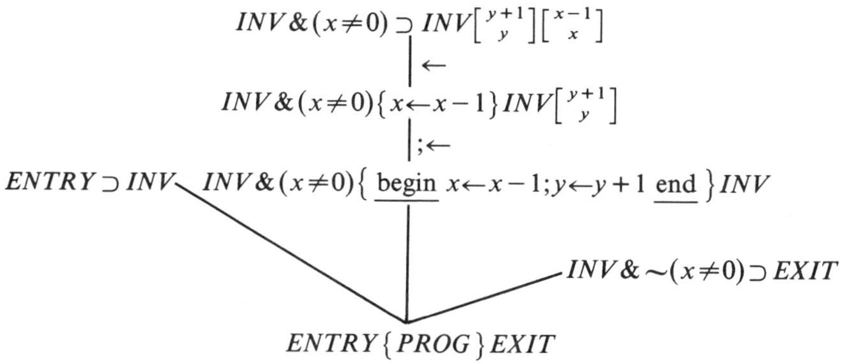

Verification Conditions

VC #1: $ENTRY\supset INV$

$$(x=x_0)\&(y=y_0)\supset(x+y=x_0+y_0)$$

VC #2: $INV\&(x\neq 0)\supset INV\left[{}^{y+1}_{y}\right]\left[{}^{x-1}_{x}\right]$

$$(x+y=x_0+y_0)\&(x\neq 0)\supset((x-1)+(y+1)=x_0+y_0)$$

VC #3: $INV\&\sim(x\neq 0)\supset EXIT$

$$(x+y=x_0+y_0)\&\sim(x\neq 0)\supset y=x_0+y_0$$

FIGURE 6.5.2

verification conditions and the subgoal

$$INV\&(x\neq 0)\{\text{ begin } x\leftarrow x-1; y\leftarrow y+1 \text{ end }\}INV$$

We applied the derived rule ;← to this subgoal, yielding

$$INV\&(x\neq 0)\{x\leftarrow x-1\}INV\left[{}^{y+1}_{y}\right]$$

We then used the assignment rule to get the verification condition

$$INV\&(x\neq 0)\supset INV\left[{}^{y+1}_{y}\right]\left[{}^{x-1}_{x}\right]$$

We then expanded the abbreviations and made the indicated substitutions to get the three verification conditions, as shown at the bottom of Figure 6.5.2. Each of the verification conditions is true in $\mathfrak{N}$, so the original assertion

$$
\begin{array}{l}
(x = x_0) \& (y = y_0) \\
\{\underline{\text{while}}\ x \neq 0\ \underline{\text{inv}}\ x + y = x_0 + y_0 \\
\qquad \underline{\text{do}}\ \underline{\text{begin}}\ x \leftarrow x - 1; \\
\qquad\qquad y \leftarrow y + 1 \\
\qquad \underline{\text{end}}\} \\
(y = x_0 + y_0)
\end{array}
$$

is true in $\mathfrak{N}$.

What is going on here? Intuitively, the program loops, decrementing x and incrementing y, until x reaches 0. So y is incremented precisely x_0 times, driving y from y_0 to $y_0 + x_0$, as desired. On the other hand, it is in general a tricky business to count the number of times a loop is executed.[16] Since x is decreased by one every time y is increased by one, execution of the statement

$$\underline{\text{begin}}\ x \leftarrow x - 1; y \leftarrow y + 1\ \underline{\text{end}}$$

leaves the value of $x + y$ unchanged; if we think of the registers x and y as containing pebbles, this statement moves one pebble from x to y. So we choose for our invariant $x + y = x_0 + y_0$, that is, "$x + y$ is the answer." Then the body of the loop preserves the invariant (VC #2), and if $x = 0$, then $y = x_0 + y_0$ (VC #3).

This addition program should have been simple.[17] Let us try a slightly more interesting program:

$$
\begin{array}{l}
\underline{\text{begin}}\ z \leftarrow 0; \\
\quad \underline{\text{while}}\ y \neq 0\ \underline{\text{do}}\ \underline{\text{begin}} \\
\qquad z \leftarrow z + x; \\
\qquad y \leftarrow y - 1 \\
\qquad \underline{\text{end}} \\
\underline{\text{end}}
\end{array}
$$

This is a program for multiplication by repeated addition. We accumulate the answer in a register z. We start it at 0, and we add x to it y times, controlling the loop by counting down y. What is the quantity that is left invariant here? Some thought reveals that execution of

$$\underline{\text{begin}}\ z \leftarrow z + x; y \leftarrow y - 1\ \underline{\text{end}}$$

leaves $x*y + z$ invariant. If the new values of z and y are denoted z' and y', respectively, we have $z' = z + x$ and $y' = y - 1$, so

$$x*y' + z' = x*(y-1) + (z+x) = x*y - x + z + x = x*y + z$$

[16]As every programmer has discovered at some time in her life!

[17]In fact, the "explanation" was probably more confusing than the program (but remember the first epigraph, at the front of this book!).

Hence we choose for our invariant $x*y+z=x_0*y_0$, and our assertion is

$$
\begin{array}{l}
(x=x_0)\&(y=y_0)\\
\{\underline{\text{begin}}\; z\leftarrow 0;\\
\qquad \underline{\text{while}}\; y\neq 0\; \underline{\text{inv}}\; x*y+z=x_0*y_0\; \underline{\text{do}}\; \underline{\text{begin}}\\
\qquad\qquad z\leftarrow z+x;\\
\qquad\qquad y\leftarrow y-1\\
\qquad\qquad \underline{\text{end}}\\
\;\underline{\text{end}}\}\\
(z=x_0*y_0)
\end{array}
$$

The proof of this assertion in system *H* is shown in Figure 6.5.3. Initializing z to 0 makes the invariant true the first time the loop is reached

FIGURE 6.5.3

Abbreviations

ENTRY	is	$(x=x_0)\&(y=y_0)$
EXIT	is	$z=x_0*y_0$
INV	is	$x*y+z=x_0*y_0$
S_1	is	$\underline{\text{begin}}\; z\leftarrow z+x; y\leftarrow y-1\; \underline{\text{end}}$
S_2	is	$\underline{\text{while}}\; y\neq 0\; \underline{\text{inv}}\; INV\; \underline{\text{do}}\; S_1$

$$INV\&(y\neq 0)\supset INV\left[{}^{y-1}_{\;y}\right]\left[{}^{z+x}_{\;z}\right]$$

$\leftarrow$

$$INV\&(y\neq 0)\{z\leftarrow z+x\}INV\left[{}^{y-1}_{\;y}\right]$$

$;\leftarrow$

$$INV\&(y\neq 0)\{S_1\}INV$$

$$ENTRY\&(z=0)\supset INV \qquad \text{while} \qquad INV\&\sim(y\neq 0)\supset EXIT$$

$$ENTRY\&(z=0)\{S_2\}EXIT$$

init

$$ENTRY\{\;\underline{\text{begin}}\; z\leftarrow 0; S_2\; \underline{\text{end}}\;\}EXIT$$

Verification Conditions

VC #1: $ENTRY\&(x=0)\supset INV$

$$(x=x_0)\&(y=y_0)\&(z=0)\supset(x*y+z=x_0*y_0)$$

VC #2: $INV\&(y\neq 0)\supset INV\left[{}^{y-1}_{\;y}\right]\left[{}^{z+x}_{\;z}\right]$

$$(x*y+z=x_0*y_0)\&(y\neq 0)\supset(x*(y-1)+(z+x)=x_0*y_0)$$

VC #3: $INV\&\sim(y\neq 0)\supset EXIT$

$$(x*y+z=x_0*y_0)\&\sim(y\neq 0)\supset(z=x_0*y_0)$$

(VC #1); the loop body preserves the invariant (VC #2), and if the invariant holds and we exit (that is, $y=0$), then the answer is in z (VC #3). Each of the verification conditions is true in $\mathfrak{N}$, so the assertion is true in $\mathfrak{N}$.

Now we reach the point that gives the method of invariants its real power: if we replace begin $z \leftarrow z+x; y \leftarrow y-1$ end by any other statement S_1 such that $INV \& (y \neq 0)\{S_1\}INV$ is true in $\mathfrak{N}$, the program would still work. For example, we could exchange the order of the assignments with confidence. There are, however, quite different manipulations that preserve the value of $x*y+z$. One that comes to mind is this: divide y by two and multiply x by two. This leaves $x*y$ unchanged, and so it also leaves the value of $x*y+z$ unchanged.

This sounds promising, since this drives y a lot farther toward zero than the old S_1, and our object is to get y down to zero so we can get out of the loop. Furthermore, if we are on a binary computer, the division and multiplication by two can be done by shifting, so we are not "cheating,"[18] at least if y is even. If y is odd, we can fall back on our old technique of just decrementing y. These considerations lead to the following program, here shown inside its assertion:

```
(x = x0) & (y = y0)
{begin z←0;
     while y≠0
          inv x*y + z = x0*y0
          do if even(y) then begin y←y/2; x←x*2 end
                        else begin z←z + x; y←y − 1 end
 end}
(z = x0*y0)
```

The proof of this assertion is shown in Figure 6.5.4. Here we have numbered all of the assertions and verification conditions that arose in the proof and put only the numbers in the tree. This makes the proof still easier to write and to read.

The clever programmer will probably object to this program. This program has a wasted test in it, she will say. If y is nonzero and even, then dividing y by two leaves y nonzero, so we are guaranteed to go around the loop at least once more. Hence the $y \neq 0$ test is wasted.

[18] As we would be if we wrote something like

```
begin z←0;
      while y≠0 do begin
            z←x*y;
            y←0
            end
end
```

But notice that magic word "leaves"! What our clever programmer has noticed is that $y \neq 0$ is an invariant:

$$\text{even}(y)\,\&\,(y \neq 0)\{\ \underline{\text{begin}}\ y \leftarrow y/2; x \leftarrow x*2\ \underline{\text{end}}\ \}(y \neq 0)$$

is true in $\mathfrak{N}$. So we can safely go around the "even" part of the loop as long as we can; when y turns odd, it is guaranteed to still be nonzero, so we can proceed directly to the "odd" branch. These considerations lead us to the following program[19]:

```
begin z←0;
    while y≠0 do
        begin
            while even(y) do
                begin y←y/2; x←x*2 end;
            begin z←z+x; y←y−1 end
        end
end
```

We now insert the invariants and the entry/exit conditions to get the assertion we need to prove:

```
(x = x0) & (y = y0)
{begin z←0;
    while y≠0 inv (x*y + z = x0*y0) do
        begin
            while even(y) inv (x*y + z = x0*y0) & (y≠0)
                do begin y←y/2; x←x*2 end;
            begin z←z+x; y←y−1 end
        end
end}  (z = x0*y0)
```

The proof is shown in Figure 6.5.5. Note that at node 4 we had a compound statement consisting of a while statement followed by a compound statement. None of the derived rules from Section 4 were applicable, so we had to use a bit of ingenuity to come up with an intermediate formula Q. We choose $INV2$ for Q, since we had already shown that

$$\mathfrak{N} \models (x*y + z = x_0*y_0)\,\&\,(y \neq 0)\{\ \underline{\text{begin}}\ z \leftarrow z + x; y \leftarrow y - 1\ \underline{\text{end}}\ \}$$
$$(x*y + z = x_0*y_0)$$

We get six verification conditions, all true in $\mathfrak{N}$, so the assertion is true in $\mathfrak{N}$.

It is also worth mentioning that once we had analyzed our programs in terms of invariants, the actual proofs in system *H* were fairly dull; they

[19]Notice we could have deleted one begin...end pair if we had allowed more than two statements in a compound statement.

Abbreviations

ENTRY	is	$(x = x_0) \& (y = y_0)$
EXIT	is	$z = x_0 * y_0$
INV	is	$x * y + z = x_0 * y_0$
S_1	is	<u>begin</u> $y \leftarrow y/2; x \leftarrow x*2$ <u>end</u>
S_2	is	<u>begin</u> $z \leftarrow z + x; y \leftarrow y - 1$ <u>end</u>
S_3	is	<u>while</u> $y \neq 0$ <u>inv</u> *INV* <u>do</u> <u>if</u> even(y) <u>then</u> S_1 <u>else</u> S_2

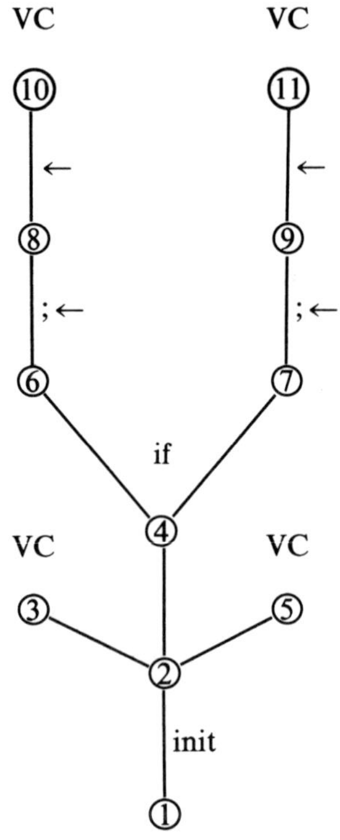

FIGURE 6.5.4

1. $ENTRY\{\underline{\text{begin}}\, z \leftarrow 0;\, S_3\, \underline{\text{end}}\}\, EXIT$
2. $ENTRY \,\&\, (z=0)\{S_3\}\, EXIT$
3. $ENTRY \,\&\, (z=0) \supset INV$ VC #1
4. $INV \,\&\, (y \neq 0)\{\underline{\text{if}}\ \text{even}(y)\ \underline{\text{then}}\ S_1\ \underline{\text{else}}\ S_2\}\, INV$
5. $INV \,\&\, \sim(y \neq 0) \supset EXIT$ VC #2
6. $INV \,\&\, (y \neq 0) \,\&\, \text{even}(y)\{S_1\}\, INV$
7. $INV \,\&\, (y \neq 0) \,\&\, \sim\text{even}(y)\{S_2\}\, INV$
8. $INV \,\&\, (y \neq 0) \,\&\, \text{even}(y)\{y \leftarrow y/2\}\, INV[{}^{x*2}_{\ x}]$
9. $INV \,\&\, (y \neq 0) \,\&\, \sim\text{even}(y)\{z \leftarrow z + x\}\, INV[{}^{y-1}_{\ y}]$
10. $INV \,\&\, (y \neq 0) \,\&\, \text{even}(y) \supset INV[{}^{x*2}_{\ x}][{}^{y/2}_{\ y}]$ VC #3
11. $INV \,\&\, (y \neq 0) \,\&\, \sim\text{even}(y) \supset INV[{}^{y-1}_{\ y}][{}^{z+x}_{\ z}]$ VC #4

Verification Conditions

VC #1: $(x = x_0) \,\&\, (y = y_0) \,\&\, (z = 0) \supset (x*y + z = x_0*y_0)$

VC #2: $(x*y + z = x_0*y_0) \,\&\, \sim(y \neq 0) \supset (z = x_0*y_0)$

VC #3: $(x*y + z = x_0*y_0) \,\&\, (y \neq 0) \,\&\, \text{even}(y) \supset ((x*2)*(y/2) + z = x_0*y_0)$

VC #4:

$$(x*y + z = x_0*y_0) \,\&\, (y \neq 0) \,\&\, \sim\text{even}(y) \supset (x*(y-1) + (z + x) = x_0 + y_0)$$

FIGURE 6.5.4 (*continued*)

Abbreviations

ENTRY	is	$(x = x_0) \& (y = y_0)$
EXIT	is	$z = x_0 * y_0$
*INV*1	is	$x * y + z = x_0 * y_0$
*INV*2	is	$INV1 \& (y \neq 0)$
S_1	is	<u>begin</u> $y \leftarrow y/2; x \leftarrow x*2$ <u>end</u>
S_2	is	<u>begin</u> $z \leftarrow z + x; y \leftarrow y - 1$ <u>end</u>
S_3	is	<u>while</u> $y \neq 0$ <u>inv</u> *INV*1 <u>do</u> <u>begin</u> <u>while</u> even(y) <u>inv</u> *INV*2 <u>do</u> S_1; S_2 <u>end</u>

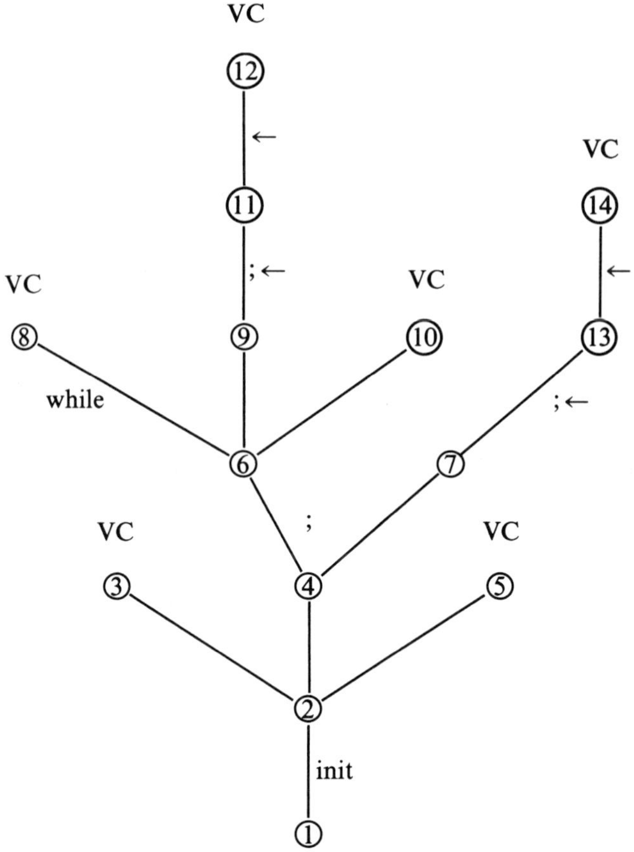

FIGURE 6.5.5

1. $ENTRY\{\underline{\text{begin}}\, z \leftarrow 0;\, S_3\, \underline{\text{end}}\}\, EXIT$
2. $ENTRY \,\&\, (z=0)\{S_3\}\, EXIT$
3. $ENTRY \,\&\, (z=0) \supset INV1$ — VC #1
4. $INV1 \,\&\, (y \neq 0)\{\underline{\text{begin}}\ \underline{\text{while}}\ \text{even}(y)\ \underline{\text{inv}}\ INV2\ \underline{\text{do}}\ S_1;\, S_2\, \underline{\text{end}}\}\, INV1$
5. $INV1 \,\&\, \sim(y \neq 0) \supset EXIT$ — VC #2
6. $INV1 \,\&\, (y \neq 0)\{\underline{\text{while}}\ \text{even}(y)\ \underline{\text{inv}}\ INV2\ \underline{\text{do}}\ S_1\}\, INV2$
7. $INV2\{S_2\}\, INV1$
8. $INV1 \,\&\, (y \neq 0) \supset INV2$ — VC #3
9. $INV2 \,\&\, \text{even}(y)\{S_1\}\, INV2$
10. $INV2 \,\&\, \sim\text{even}(y) \supset INV2$ — VC #4
11. $INV2 \,\&\, \text{even}(y)\{y \leftarrow y/2\}\, INV2[^{x*2}_{x}]$
12. $INV2 \,\&\, \text{even}(y) \supset INV2[^{x*2}_{x}][^{y/2}_{y}]$ — VC #5
13. $INV2\{z \leftarrow z + x\}\, INV1[^{y-1}_{y}]$
14. $INV2 \supset INV1[^{y-1}_{y}][^{z+x}_{z}]$ — VC #6

Verification Conditions

VC #1: $(x = x_0) \,\&\, (y = y_0) \,\&\, (z = 0) \supset (x*y + z = x_0*y_0)$

VC #2: $(x*y + z = x_0*y_0) \,\&\, \sim(y \neq 0) \supset (z = x_0*y_0)$

VC #3: $(x*y + z = x_0*y_0) \,\&\, (y \neq 0) \supset (x*y + z = x_0*y_0) \,\&\, (y \neq 0)$

VC #4: $(x*y + z = x_0*y_0) \,\&\, (y \neq 0) \,\&\, \sim\text{even}(y) \supset$
$(x*y + z = x_0*y_0) \,\&\, (y \neq 0)$

VC #5: $(x*y + z = x_0*y_0) \,\&\, (y \neq 0) \,\&\, \text{even}(y) \supset$
$((x*2)*(y/2) + z = x_0*y_0) \,\&\, (y/2 \neq 0)$

VC #6: $(x*y + z = x_0*y_0) \,\&\, (y \neq 0) \supset (x*(y-1) + (z + x) = x_0*y_0)$

FIGURE 6.5.5 (*continued*)

merely confirmed that everything fits together the way we thought. In the remainder of this chapter, we shall concentrate on this informal stage of analysis to illustrate how invariants can be used to help *write* programs.

EXERCISES 6.5

Prove the following assertions in system H, in the data type $\mathfrak{N}$ (with division by 2 and the test for evenness allowed):

1. $(x=x_0)\&(y=y_0)$
$\{\underline{\text{while}}\ (x\neq 0)\&(y\neq 0)\ \underline{\text{inv}}\ x\dot{-}y=x_0\dot{-}y_0$
$\quad \underline{\text{do}}\ \underline{\text{begin}}\ x\leftarrow x-1; y\leftarrow y-1\ \underline{\text{end}}\}\ x=x_0\dot{-}y_0$

2. $(x=x_0)\&(y=y_0)$
$\{\underline{\text{begin}}\ \ z\leftarrow 1;$
$\quad \underline{\text{while}}\ y\neq 0\ \underline{\text{inv}}\ z*x^y=x_0^{y_0}\ \underline{\text{do}}\ \underline{\text{begin}}$
$\quad\quad z\leftarrow z*x;$
$\quad\quad y\leftarrow y-1$
$\quad\quad \underline{\text{end}}$
$\underline{\text{end}}\}$
$z=x_0^{y_0}$

3. $(x=x_0)\&(y=y_0)$
$\{\underline{\text{begin}}$
$\quad \underline{\text{begin}}\ q\leftarrow 0; r\leftarrow x\ \underline{\text{end}};$
$\quad \underline{\text{while}}\ r\geqslant y\ \underline{\text{inv}}\ (q*y+r=x_0)\ \underline{\text{do}}\ \underline{\text{begin}}$
$\quad\quad q\leftarrow q+1;$
$\quad\quad r\leftarrow r-y$
$\quad\quad \underline{\text{end}}$
$\underline{\text{end}}\}$
$(x_0=q*y+r)\&(r<y)$

4. $(x=x_0)$
$\{\underline{\text{begin}}\ z\leftarrow 1;$
$\quad \underline{\text{while}}\ x\neq 1\ \underline{\text{inv}}\ z*x!=x_0!\ \underline{\text{do}}\ \underline{\text{begin}}$
$\quad\quad z\leftarrow z*x;$
$\quad\quad x\leftarrow x-1$
$\quad\quad \underline{\text{end}}$
$\underline{\text{end}}\}$
$z=x_0!$

5. $(x=x_0)\&(y=y_0)$
$\{\underline{\text{begin}}\ \ z\leftarrow 1;$
$\quad \underline{\text{while}}\ y\neq 0\ \underline{\text{inv}}\ z*x^y=x_0^{y_0}\ \underline{\text{do}}$
$\quad\quad \underline{\text{if}}\ \text{even}\,(y)\ \underline{\text{then}}\ \underline{\text{begin}}\ y\leftarrow y/2; x\leftarrow x*x\ \underline{\text{end}}$
$\quad\quad\quad \underline{\text{else}}\ \underline{\text{begin}}\ y\leftarrow y-1; z\leftarrow z*x\ \underline{\text{end}}$
$\underline{\text{end}}\}$
$(z=x_0^{y_0})$

6. $(x=x_0)\&(y=y_0)$

```
{begin z←1;
    while y≠0 inv z∗x^y = x0^y0 do
        begin
            while even(y) inv (z∗x^y = x0^y0)&(y≠0)
                do begin y←y/2; x←x∗x end;
            begin z←z∗x; y←y−1 end
        end
 end} z = x0^y0
```

6.6 WRITING PROGRAMS USING INVARIANTS

In this section we shall see how it is possible to use loop invariants to write programs. We shall construct an informal correctness proof *as we write* the program, so that the program should be correct when it is completed.[20] We shall do two examples in some detail to show how to construct invariants and programs together.

Let us first see how having an assertion in mind simplifies writing a program. If we have formulas P and Q and we are to write a program S such that $P\{S\}Q$ is true, then the presence of the formula P is a great help: It tells us that program S need only work properly on input that satisfies P. P tells us what assumptions S can make about the state of the world when it starts. In other words, P tells us what constitutes "good data" for S. Thus the goal $INV\&B\{S\}INV$ for the body of the loop says: If S is happy when it starts, it should make things happy for the next time around.

One elementary notion of "good data" is "within range." Let us see how that helps us write an integer square root program. The desired assertion is

$$x \geqslant 0\{S\}(z^2 \leqslant x)\&((z+1)^2 > x)$$

We propose to do this by a *linear search*. That is, we shall start z at 0 and increment it until we get to the answer. So z is "within range" iff $(0 \leqslant z)\&(z^2 \leqslant x)$. In fact, let us restrict ourselves to the data type $\mathfrak{N}$ of nonnegative integers, so that $0 \leqslant z$ is always true. Then we propose $(z^2 \leqslant x)$ as our invariant.

Now we also have a good idea about what the loop body will look like, since we already said we were going to increment z repeatedly. So now our program looks like

```
begin z←0;
    while B inv (z^2 ≤ x) do z←z+1;
    maybe some cleanup[21]
end
```

[20]Now, this is not really so different from the normal way we write programs. After all, we do not write programs unless we have a good reason to believe they work. What the method of invariants supplies is a better way to analyze whether programs work.

[21]From now on we shall allow a compound statement to have the form:

$$\underline{\text{begin}}\ S_1; \ldots ; S_n\ \underline{\text{end}}$$

We now have to propose a candidate for B. One thing that B has to do is to prevent the $z \leftarrow z+1$ from driving z out of bounds. That is, we have the following scheme:

```
begin initialize;
      while it's safe do take a step;
      clean up
end
```

where "it's safe" means "taking a step will not cause the invariant to be violated."[22] So B has to "look ahead."[23] In our example, we propose $(z+1)^2 \leqslant x$ for B, giving

```
begin z←0;
      while (z+1)² ⩽ x inv (z² ⩽ x) do z←z+1;
      clean up
end
```

Now $\mathfrak{N} \models (z^2 \leqslant x) \& ((z+1)^2 \leqslant x)\{z \leftarrow z+1\}(z^2 \leqslant x)$, so the invariant really is preserved by the loop. Now all that is left is to clean up. After the loop exits, we know that the invariant $z^2 \leqslant x$ is still true and that the test $(z+1)^2 \leqslant x$ is false, that is, $(z+1)^2 > x$. But that is just the desired output condition! So we are done. The completed program is

```
begin z←0;
      while (z+1)² ⩽ x
            inv z² ⩽ x
            do z←z+1
end
```

This is a correct program, but it is not a very good program. Each pass through the loop involves a squaring operation, which is slow on most computers. We can eliminate the squaring operation by introducing a new variable u and adding $u=(z+1)^2$ to the invariant. So our loop becomes

```
while u ⩽ x
      inv (z² ⩽ x) & (u = (z+1)²)
      do begin
             z←z+1;
             adjust u
         end
```

If we can write the loop body to maintain the invariant $u=(z+1)^2$, then the new version of the test will work exactly the way the old one did. So to

[22]Another phrasing is: B has to *guard* against violation of the invariant. This was the role that the even (y) test played in the nested-loop multiplication program.

[23]Cf. the role of $Q[^t_v]$ in the assignment rule.

maintain the invariant we do some elementary algebra:

$$\begin{aligned} new\,z &= z+1 \\ new\,u &= (new\,z+1)^2 \\ &= new\,z^2+2\cdot new\,z+1 \\ &= (z+1)^2+2\cdot new\,z+1 \\ &= u+2\cdot new\,z+1 \end{aligned}$$

Now the loop reads

```
while u ≤ x
    inv (z² ≤ x)&(u=(z+1)²)
    do begin
            z←z+1;
            u←u+2z+1
       end
```

We wrote the loop body to maintain the invariant, so (by golly) the invariant is maintained! So at the end of the loop we still have $(z^2 \leqslant x)\&((z+1)^2 > x)$ since $u > x$ and $u=(z+1)^2$. Furthermore, the expensive squaring operation has been replaced by additions and multiplications.[24] Now all we have to do is initialize u so that the invariant will be true the first time the loop is reached. We start with $z=0$, so we should set $u=(z+1)^2=(0+1)^2=1$. The program is now

```
begin z←0; u←1;
    while u ≤ x
        inv (z² ≤ x)&(u=(z+1)²)
        do begin z←z+1; u←u+2z+1 end
end
```

Notice that there has been a subtle change in our thinking. We originally thought of an invariant as specifying the data that made a program happy. Now we think of a program as a device to keep an invariant happy.[25] This symbiosis may be expressed by the acronymic slogan GIGO:[26] Good stuff In, Good stuff Out.

Let us try another example. We are given two real numbers x and y with $0 \leqslant x < y \leqslant 1$, and we are to find their quotient z within a tolerance *tol*, without, of course, using division. More precisely, we have to find S such

[24]Such a replacement is called a "reduction in strength," and can be performed by an optimizing compiler to replace multiplications in a loop by additions. This situation typically arises in array references. Dijkstra uses the term "moving an invariant outside a repetitive construct" for a similar transformation [Dijkstra 76].

[25]Now be honest: would you have initialized u to zero without knowing the invariant?

[26]Formerly "garbage in—garbage out."

that (see Figure 6.6.1)

$$0 \leqslant x < y \leqslant 1\{S\} z \leqslant x/y < z + tol$$

Our strategy is depicted by the following scheme:

```
begin
      make a guess;
      while not done
               inv this is a reasonable guess
               do improve your guess
end
```

In order to tell whether we are done, we must know whether we have gotten within the tolerance *tol*. To that end, we introduce a new variable *d*, which is to hold the known accuracy of our current guess. That is, our guess is reasonable if it satisfies

$$z \leqslant x/y < z + d$$

which becomes the loop invariant.

Since $0 \leqslant x < y$ we can initialize with $x = 0$ and $d = 1$, getting

```
begin z←0; d←1;
      while d > tol
               inv z ≤ x/y < z + d
               do improve z and d
end
```

Since we know that the answer is between z and $z+d$, we can try the technique of *binary search*: we try $z + \frac{1}{2}d$. If $z + \frac{1}{2}d > x/y$, then $z \leqslant x/y < z + \frac{1}{2}d$; otherwise, $z + \frac{1}{2}d \leqslant x/y < z + d$. Either way we know the quotient to within $\frac{1}{2}d$. So[27]

```
begin z←0; d←1;
      while d > tol
           inv z ≤ x/y < z + d
           do begin if z + ½d > x/y then z←z
                                    else z←z + ½d;
                    d←½d
              end
end
```

[27]We could avoid the assignment $z \leftarrow z$ if we had had the single-branch conditional: if B then S.

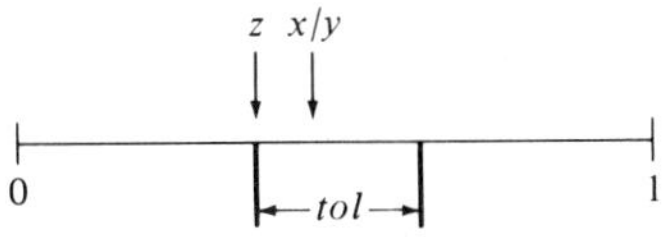

FIGURE 6.6.1 Desired situation at termination of division algorithm.

Of course this program cheats by using division, so let us multiply both sides of the test by y:

```
begin z←0; d←1;
    while d > tol
        inv z ≤ x/y < z + d
        do begin if zy + ½dy > x then z←z
                                 else z←z + ½d;
              d←½d
           end
end
```

Now we get rid of the multiplications by introducing new variables u and v subject to the invariants $u = zy$ and $v = \frac{1}{2}dy$. Now we have to update u and v, depending on the test:

Case I $(u + v > x)$:

$$\begin{aligned} new\, d &= \tfrac{1}{2}d \\ new\, z &= z \\ new\, u &= new\, z \cdot y = z \cdot y = u \\ new\, v &= \tfrac{1}{2}(new\, d) \cdot y = \tfrac{1}{2}\left(\tfrac{1}{2}d\right)y = \tfrac{1}{2}v \end{aligned}$$

Case II $(u + v \leqslant x)$:

$$\begin{aligned} new\, d &= \tfrac{1}{2}d \\ new\, z &= z + \tfrac{1}{2}d = z + new\, d \\ new\, u &= new\, z \cdot y = \left(z + \tfrac{1}{2}d\right)y = zy + \tfrac{1}{2}dy = u + v \\ new\, v &= \tfrac{1}{2}(new\, d) \cdot y = \tfrac{1}{2}\left(\tfrac{1}{2}d\right)y = \tfrac{1}{2}v \end{aligned}$$

This yields the following program:

```
begin z←0; d←1; u←0; v←½y;
    while d > tol
        inv (z ≤ x/y < z + d) & (u = zy) & (v = ½dy)
        do if u + v > x
            then begin d←½d; v←½v end
            else begin d←½d; z←z + d; u←u + v; v←½v end
end
```

(Notice that here, unlike most of our examples, the order of the assignments makes a difference.)

We now have a real number division program that uses only addition, subtraction, and division by two. Thus it is a good candidate for use in a binary machine for dividing the fractional parts of floating-point numbers. This algorithm is called Wensley's algorithm [Wensley 58].

Variables like u and v are called *trailer variables*. They are used to hold information that could be recomputed at every step but is rather maintained and updated each time through the loop.[28] We get the picture of the main program variables taking a step and the trailer variables scurrying around attempting to "catch up," where "catching up" means making the invariant happy. This notation of *stepping under an invariant* is crucial to an understanding of the complex algorithms typical in more advanced data structures.[29]

EXERCISES 6.6

1. Prove in system H the square-root program on

   (a) page 178
   (b) page 179

In the following exercises, use the method of invariants, as discussed in this section.

2. Modify the square root program on page 179 by adding to the invariant the condition "$v = 2z + 1$."

3. In the division program on page 181 eliminate the addition in the test.

4. Write a program over the data type $\mathfrak{N}$, which, given an integer x, finds the number of "1's" in the binary representation of x.

5. Write a program over the data type $\mathfrak{N}$, which finds the integer quotient of two positive integers using binary search.

6. Write a program that finds the square root of a real number $0 \leqslant x < 1$ to a specified tolerance, using binary search.

*7. Write a program that, given a list (Section 4.1), counts the number of atoms in the list. (Assume that integer arithmetic is available.)

[28]See Chapter 3, footnote 30.
[29]For a discussion of these ideas, see [Wirth 73] or [Wirth 74].

6.7 HANDLING ARRAYS

The programs we have discussed so far are, of course, somewhat unrealistic. In order to deal with more realistic programming problems we must at least be able to talk about arrays. Once arrays are available, we can discuss searching, sorting, rearranging, and similar tasks that constitute a substantial portion of the programmer's bag of tricks. With arrays, we can introduce a bug by forgetting to look at an entry, looking at an entry twice, or accessing a nonexistent entry. In this section we shall show how the technique of writing programs with invariants can help avoid these common bugs.

Although it is possible to extend the formal analysis of system H to handle arrays, the resulting system is more cumbersome and is beyond the scope of this text. The informal analysis of Section 6, however, is easily extended to handle arrays.

Let us first attack the problem of finding the smallest element of an array. Let us assume we have an array A, with subscripts ranging from 1 to n, filled with real numbers. We can use the "improve your guess" scheme:

```
begin
      make a guess;
      while not done
                inv this is a reasonable guess
                do improve your guess
end
```

For our definition of "reasonable guess" we can choose

$A[min]$ is the smallest element seen so far

giving

```
begin
      make a guess;
      while not done
        inv A[min] is the smallest element seen so far
        do begin
                look at a new element;
                update min
           end
end
```

Now we can write the invariant as a formula

$$INV1: \quad (\forall i)\{(1 \leqslant i \leqslant sofar) \supset (A[min] \leqslant A[i])\}$$

involving the variable *sofar*, which means the step "look at a new element" means "$sofar \leftarrow sofar + 1$," and "not done" means "$sofar \neq n$," giving

```
begin
      make a guess;
      while sofar≠n
            inv INV1
            do begin
                     sofar←sofar + 1;
                     update min[30]
               end
end
```

Now we know that when the loop exits we shall have $INV1$ true and $sofar = n$, so

$$(\forall i)\{(1 \leqslant i \leqslant n) \supset (A[min] \leqslant A[i])\}$$

which is the desired postcondition.[31]

Now all we have to do is write code for "make a guess" and "update *min*" which keeps the invariant happy. But that is easy:

```
begin
      sofar←1; min←1;
      while sofar≠n
        inv INV1
        do begin
                  sofar←sofar + 1;
                  if A[sofar]<A[min] then min←sofar
                                     else min←min
           end
end
```

This is the "find the minimum" program familiar to every beginning programming student, except that by using invariants we avoided the bugs attendant upon things like DO-loop indices.

Having found the smallest element of our array, let us try sorting it. Our "improve your guess" scheme can be used, where "a reasonable guess"

[30]It is amusing that here *min* is a trailer variable, after a fashion.

[31]If we had instead written the test as while $sofar < n$ do..., we would have had to include $sofar \leqslant n$ in the invariant, in order to conclude $sofar = n$ at exit. Since it is just as easy to test for "$\neq$" as "$<$," using the "$\neq$" is better because it makes it easier to demonstrate correctness.

means "a good partial answer":

```
begin
    initialize;
    while not done
        inv this is a good partial answer
        do improve the answer
end
```

What is "a good partial answer"? One possibility is shown in Figure 6.7.1. This picture has the pleasant property that if $s=1$, then the original array is a legitimate partial answer, so we can initialize by setting s to 1. Furthermore, if $s=n$ and the array satisfies the picture, then the array is sorted. So the loop test can be while $s \neq n$ do....

We can express Figure 6.7.1 as a first-order formula as follows:

$$INV2: \quad (\forall i)(\forall j)[1 \leqslant i \leqslant j \leqslant s \supset A[i] \leqslant A[j]]$$

This leads to the following program:

```
begin
    s←1;
    while s≠n
        inv INV2
        do improve the answer
end
```

Of course, we don't know that we can "improve the answer." A trivial program that maintains $INV2$ is "$s \leftarrow s$." But that can hardly be called progress. Since we start with $s=1$ and finish with $s=n$, a candidate for "improve the answer" is

```
begin
    s←s+1;
    rearrange A[1] through A[s] to make INV2 true again
end
```

This leads to the vision of a "wave of sortedness" sweeping through the array.[32]

Now we must figure out how to do the rearrangement. This goal is shown in Figure 6.7.2. Here $A[s]$ is inserted at location $p+1$, and the contents of $A[p+1]$ through $A[s-1]$ are shifted right one place. In order to do this, we have to find p such that $A[p] \leqslant x \leqslant A[p+1]$. Since $A[1]$ through $A[s-1]$ are sorted,[33] this can be done by finding the *largest* p such

[32]Like light, data is best viewed sometimes as a particle and sometimes as a wave.
[33]Remember, we have just incremented s.

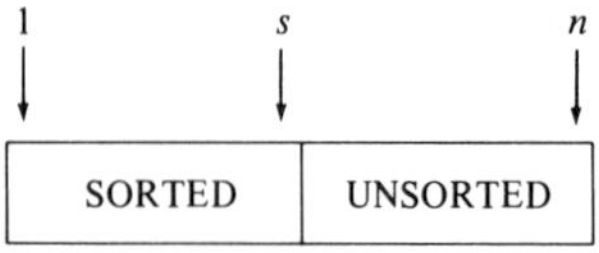

FIGURE 6.7.1

that $A[p] \leqslant A[s]$. This can be done by a routine linear search:

```
begin
    p←s−1;
    while A[p]>A[s] do p←p−1
end
```

Unfortunately, this will do nasty things if $A[s]$ is smaller than any of $A[1]$ through $A[s-1]$. We could try to solve this by using a more complicated loop test, such as

```
begin
    p←s−1;
    while (p>0)&(A[p]>A[s]) do p←p−1
end
```

but this will fail unless we make some special rules about not evaluating both sides of the "&." A more elegant solution is to introduce a special array element, $A[0]$, and load it with $A[s]$ so that the loop is guaranteed to exit before anything unpleasant happens. Since $A[s]$ is stored away, we can safely shift the data upward as we go.

```
begin
    A[0]←A[s];
    p←s−1;
    while A[0]<A[p]
          inv (∀i)[p<i≤s ⊃ A[i]>A[0]]
          do begin
                    A[p+1]←A[p];
                    p←p−1
             end;
    A[p+1]←A[0]
end
```

Notice how the invariant changed to incorporate the way the data was

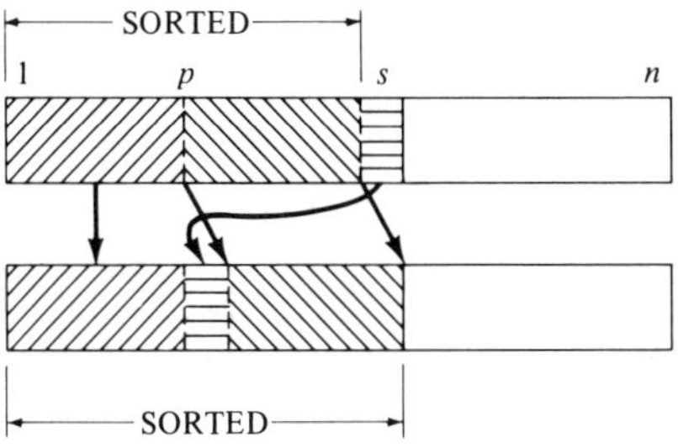

FIGURE 6.7.2

shifted.[34] We can now write the entire sorting program[35]:

```
begin
      s←1;
      while s≠n
            inv INV2
            do begin
                  s←s+1;
                  A[0]←A[s];
                  p←s−1;
                  while A[0]<A[p]
                        inv (∀i)[p<i⩽s⊃A[i]>A[0]]
                        do begin
                              A[p+1]←A[p];
                              p←p−1
                           end;
                  A[p+1]←A[0]
               end
end
```

Let us try one more example to see how a good choice of invariant can help solve a hard-looking problem. Assume we are given our array A of reals and a real number x. Our task is to rearrange the elements of A so that all the elements which are less than x are to the left of all the elements

[34]This trick is called using a *sentinel* [Wirth 74].

[35]The sequence of assignments between the outer and inner loops may now be simplified to

$$p \leftarrow s;\ s \leftarrow s+1;\ A[0] \leftarrow A[s]$$

thus saving a grand total of one subtraction. (Consider the room for error if we had attempted this optimization earlier!)

which are greater than x (Figure 6.7.3). That is,

$$(\forall i)(\forall j)[(1 \leqslant i \leqslant n) \& (1 \leqslant j \leqslant n) \& (A[i] \leqslant x \leqslant A[j]) \supset (i \leqslant j)]$$

This is called the partition problem.

If we think about data as waves, we might come up with the following notion of a partial solution:

$$INV3: \quad (\forall i)[(1 \leqslant i < lo) \supset (A[i] \leqslant x)] \& (\forall j)[(hi < j \leqslant n) \supset (x \leqslant A[j])]$$

which formalizes the picture of Figure 6.7.4 (which probably came before the formula).

Here *lo* and *hi* should get squeezed together until the unknown region is empty. Notice that it is possible for $INV3 \& (lo > hi)$ to hold (the elements in the overlap must all be equal to x). Hence we can play safe and implement "not done" as "$lo \leqslant hi$" (if $lo = hi$ the unknown region has one element). This gives

```
begin
      lo←1; hi←n;
      while  lo ⩽ hi
             inv INV3 (Figure 6.7.4)
             do squeeze lo and hi
end
```

If $A[lo] \leqslant x$, we can increment *lo* "for free"; a similar case arises if $A[hi] \geqslant x$. In the remaining case $A[lo] > x$ and $A[hi] < x$, so we can exchange them and advance both *lo* and *hi*! Thus:

```
begin
      lo←1; hi←n;
      while  lo ⩽ hi
             inv INV3 (Figure 6.7.4)
             do if A[lo] ⩽ x then lo←lo+1
                else if A[hi] ⩾ x then hi←hi−1
                else begin temp←A[lo];
                           A[lo]←A[hi];
                           A[hi]←temp;
                           lo←lo+1;
                           hi←hi−1
                     end
end
```

FIGURE 6.7.3

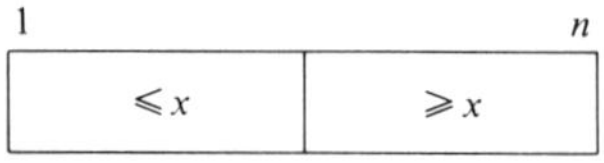

FIGURE 6.7.4

It may take considerable cleverness to arrive at a good invariant like Figure 6.7.4, but once we have a good invariant diagram, even very complicated programs reduce to mostly bookkeeping. This simplification is at the heart of most of the standard algorithms studied in data structures.

EXERCISES 6.7

Write the following programs using invariants. In each case we are working with an array A of reals, indexed from 1 to n.

1. Find the location of the largest element in A.

2. Given a real number x, find i such that $A[i]=x$ or $i=0$ if no element of A contains x.

3. Same as Exercise 2, but A is sorted. Use binary search.

4. Sort the array, strengthening $INV2$ (p. 185) to:

$$(\forall i)[1 \leqslant i \leqslant s \supset A[i] \text{ contains the } i\text{th smallest element of the array } A]$$

***5.** Given an array A as above, except that each element is either 0, 1 or 2, rearrange A so it looks like

00...00111...1122...22

with all the 0's at one end, all the 2's at the other, and all the 1's in between. *Restriction:* You may use no arrays other than A, and you may not store in A anything other than 0, 1, or 2.

6. Improve the partition program by introducing inner loops where possible.

***7.** Write a sorting program that uses the partition program.

8. A sorting program is called *stable* if it never interchanges equal elements. Analyze the sorting programs of this section and the exercises to see which ones are stable.

****9.** Extend the techniques of this chapter to handle input/output. Write a merge/update program and verify it.

**Two asterisks indicate a very difficult exercise.

References

[ACM 68]
Association for Computing Machinery
"Curriculum 68—Recommendations for Academic Programs in Computer Science—A Report of the ACM Curriculum Committee on Computer Science,"
Comm. ACM 11 (1968), 151–197.

[Baker 72]
Baker, F. T.
"System Quality Through Structured Programming,"
Proc. Fall Joint Computer Conference (1972), pp. 339–343.

[Berge 71]
Berge, C.
Principles of Combinatorics.
Academic Press, New York, 1971.

[Brainerd and Landweber 74]
Brainerd, W. S., and Landweber, L. H.
Theory of Computation.
Wiley (Interscience), New York, 1974.

[Cook 78]
Cook, S. A.
"Soundness and Completeness of an Axiom System for Program Verification,"
SIAM J. Computing 7 (1978), 70–90.

[Denning, Dennis, and Qualitz 78]
Denning, P. J., Dennis, J. B., and Qualitz, J. E.
Machines, Languages, and Computation.
Prentice-Hall, Englewood Cliffs, New Jersey, 1978.

[Dijkstra 68]
Dijkstra, E. W.
"Go to Statement Considered Harmful,"
Comm. ACM 11 (1968), 147–148.

[Dijkstra 72]
Dijkstra, E. W.
"Notes on Structured Programming,"
in Dahl, O. J., Dijkstra, E. W., and Hoare, C. A. R.,
Structured Programming. Academic Press, London, 1972.

[Dijkstra 76]
Dijkstra, E. W.
A Discipline of Programming.
Prentice-Hall, Englewood Cliffs, New Jersey, 1976.

[Floyd 67]
Floyd, R. W.
"Assigning Meanings to Programs,"
in *Proc. Symp. Appl. Math., Vol.* 19: *Mathematical Aspects of Computer Science* (J. T. Schwartz, ed.), pp. 19–32.
American Mathematical Society, Providence, Rhode Island, 1967.

[Friedman 74]
Friedman, D. P.
The Little LISPer.
Science Research Associates, Palo Alto, California, 1974.

[Griswold, Poage, and Polonsky 71]
Griswold, R. E., Poage, J. F., and Polonsky, I. P.
The SNOBOL4 Programming Language (2nd ed.).
Prentice-Hall, Englewood Cliffs, New Jersey, 1971.

[Herstein 64]
Herstein, I. N.
Topics in Algebra.
Blaisdell, Waltham, Massachusetts, 1964.

[Hoare 69]
Hoare, C. A. R.
"An Axiomatic Basis for Computer Programming,"
Comm. ACM 12 (1969), 576–580, 583.

[Hoare 72]
Hoare, C. A. R.
"Proving Correctness of Data Representations,"
Acta Informatica 1 (1972), 271–281.

[Hofstadter 79]
Hofstadter, D. R.
Gödel, Escher, Bach: An Eternal Golden Braid.
Basic Books, New York, 1979.

[Horowitz and Sahni 76]
Horowitz, E., and Sahni, S.
Fundamentals of Data Structures.
Computer Science Press, Woodland Hills, California, 1976.

[ICRS 75]
"Proceedings–1975 International Conference on Reliable Software,"
SIGPLAN Notices 10(6) (June 1975).

[Kalish and Montague 64]
Kalish, D., and Montague, R.
Techniques of Formal Reasoning.
Harcourt Brace Jovanovich, New York, 1964.

[Kleene 52]
Kleene, S. C.
Introduction to Metamathematics.
Van Nostrand, Princeton, New Jersey, 1952.

[Knuth 68]
Knuth, D. E.
The Art of Computer Programming, Vol. I: Fundamental Algorithms.
Addison–Wesley, Reading, Massachusetts, 1968 (2nd ed. 1973).

[Knuth 74]
Knuth, D. E.
"Structured Programming with Go To Statements,"
Computing Surveys 6 (1974), 261–301.

[Korfhage 74]
Korfhage, R. R.
Discrete Computational Structures.
Academic Press, New York, 1974.

[Landin 66]
Landin, P. J.
"The Next 700 Programming Languages,"
Comm. ACM 9 (1966), 157–166.

[Manna 74]
Manna, Z.
Mathematical Theory of Computation.
McGraw–Hill, New York, 1974.

[Manna and Vuillemin 72]
Manna, Z., and Vuillemin, J.
"Fixpoint Approach to the Theory of Computation,"
Comm. ACM 15 (1972), 528–536.

[Manna and Waldinger 78]
Manna, Z., and Waldinger, R.
"Is 'Sometime' Sometimes Better Than 'Always'?,"
Comm. ACM 21 (1978), 159–172.

[McCarthy et al. 65]
McCarthy, J., Abrahams, P. W., Edwards, D. J., Hart, T. P., and Levin, M. I.
LISP 1.5 Programmer's Manual.
MIT Press, Cambridge, Massachusetts, 1965.

[Morris and Wegbreit 77]
Morris, J. H. Jr., and Wegbreit, B.
"Subgoal Induction,"
Comm. ACM 20 (1977), 209–222.

[Nilsson 71]
Nilsson, N. J.
Problem-Solving Methods in Artificial Intelligence.
McGraw–Hill, New York, 1971.

[Prather 76]
Prather, R.
Discrete Mathematical Structures for Computer Science.
Houghton Mifflin, Boston, 1976.

[Pratt 75]
Pratt, T. W.
Programming Languages: Design and Implementation.
Prentice–Hall, Englewood Cliffs, New Jersey, 1975.

[Preparata and Yeh 73]
Preparata, F. P., and Yeh, R. T.
Introduction to Discrete Structures.
Addison–Wesley, Reading, Massachusetts, 1973.

[Ralston 65]
Ralston, A.
A First Course in Numerical Analysis.
McGraw–Hill, New York, 1965.

[Shoenfield 67]
Shoenfield, J. R.
Mathematical Logic.
Addison–Wesley, Reading, Massachusetts, 1967.

[Stanat and McAllister 77]
Stanat, D. F., and McAllister, D. F.
Discrete Mathematics in Computer Science.
Prentice–Hall, Englewood Cliffs, New Jersey, 1977.

[Tremblay and Manohar 75]
Tremblay, J. T., and Manohar, R. P.
Discrete Mathematical Structures with Applications to Computer Science.
McGraw–Hill, New York, 1975.

[Wand 78]
Wand, M.
"A New Incompleteness Result for Hoare's System,"
J. ACM 25 (1978), 168–175.

[Wegbreit and Spitzen 76]
Wegbreit, B., and Spitzen, J. M.
"Proving Properties of Complex Data Structures,"
J. ACM 23 (1976), 389–396.

[Wensley 58]
Wensley, J. H.
"A Class of Non-Analytical Iterative Processes,"
Computer J. 1 (1958), 163–167.

[Wilkes 66]
Wilkes, M. V.
A Short Introduction to Numerical Analysis.
Cambridge University Press, Cambridge, England, 1966.

[Wirth 73]
Wirth, N.
Systematic Programming: An Introduction.
Prentice–Hall, Englewood Cliffs, New Jersey, 1973.

[Wirth 74]
Wirth, N.
"On the Composition of Well-Structured Programs,"
Comp. Surveys 6 (1974), 247–259.

[Wirth 76]
Wirth, N.
Algorithms + Data Structures = Programs.
Prentice–Hall, Englewood Cliffs, New Jersey, 1976.

[Yousse 64]
Yousse, B. K.
Mathematical Induction.
Prentice–Hall, Englewood Cliffs, New Jersey, 1964.

Table of Upper- and Lower-Case Greek Letters

Α α	alpha	Ν ν	nu
Β β	beta	Ξ ξ	xi
Γ γ	gamma	Ο ο	omicron
Δ δ	delta	Π π	pi
Ε ϵ	epsilon	Ρ ρ	rho
Ζ ζ	zeta	Σ σ	sigma
Η η	eta	Τ τ	tau
Θ θ	theta	Υ υ	upsilon
Ι ι	iota	Φ φ	phi
Κ κ	kappa	Χ χ	chi
Λ λ	lambda	Ψ ψ	psi
Μ μ	mu	Ω ω	omega

Index